Lie Algebras and Locally Compact Groups

lie algebras and locally compact groups

[Irving Kaplansky]

The University of Chicago Press

Chicago Lectures in Mathematics Series
Irving Kaplansky, *Editor*

The Theory of Sheaves, by Richard G. Swan (1964)
Topics in Ring Theory, by I. N. Herstein (1969)
Fields and Rings, by Irving Kaplansky (1969, 2nd ed. 1972)
Infinite Abelian Group Theory, by Phillip A. Griffith (1970)
Topics in Operator Theory, by Richard Beals (1971)
Lie Algebras and Locally Compact Groups, by Irving Kaplansky (1971)
Several Complex Variables, by Raghavan Narasimhan (1971)
Torsion-Free Modules, by Eben Matlis (1972)
The Theory of Bernoulli Shifts, by Paul C. Shields (1973)
Stable Homotopy and Generalized Homology,
by J. Frank Adams (1974)

International Standard Book Number: 0-226-42453-7
Library of Congress Catalog Card Number: 76-136207

The University of Chicago Press, Chicago 60637
The University of Chicago Press, Ltd., London

Published 1971. Second impression, with additions, 1974
Printed in the United States of America

TO

TERRY MIRKIL

AND

HIDEHIKO YAMABE

IN MEMORIAM

CONTENTS

PREFACE   ix

Chapter I.   LIE ALGEBRAS

   1.   Definitions and examples   1

   2.   Solvable and nilpotent algebras   9

   3.   Semi-simple algebras   32

   4.   Cartan subalgebras   39

   5.   Transition to a geometric problem
        (characteristic 0)   47

   6.   The geometric classification   54

   7.   Transition to a geometric problem
        (characteristic p)   64

   8.   Transition to a geometric problem
        (characteristic p), continued   74

Chapter II.   THE STRUCTURE OF LOCALLY COMPACT GROUPS

   1.   NSS groups   87

   2.   Existence of one-parameter subgroups   89

   3.   Differentiable functions   97

   4.   Functions constructed from a single Q 101

   5.   Functions constructed from a sequence of Q's 104

   6.   Proof that i/n_i is bounded 109

   7.   Existence of proper differentiable functions 112

CONTENTS

8. The vector space of one-parameter subgroups 114

9. Proof that K is a neighborhood of 1 120

10. Approximation by NSS groups 131

11. Further developments 138

BIBLIOGRAPHY 143

INDEX 147

PREFACE

In the Autumn of 1957 and Winter of 1958, I presented a two-quarter course entitled "Lie algebras and Lie groups". The Autumn course was a purely algebraic account of Lie algebras. The Winter course began with the solution of Hilbert's fifth problem; this merged into an exposition of some of the foundations of Lie group theory.

During 1960 and 1961 notes on part of the course were written. Because of administrative duties, writing plans were then postponed for many years.

In 1969 I revised the 1960-1 notes on Lie algebras and, in multilith form, they were used as a partial text for a course in the Autumn of 1969. The course itself went on to other topics: principally representations and the Whitehead lemmas; this material is not reproduced here. Instead, I have added to the present account two sections (§7 and §8) on characteristic p, which were not presented in class in either version of the course. In these sections I carry out a project which I have had in mind ever since the appearance of Seligman's thesis [19]: the use of "projective" representations as a simple method of capturing more of the classical simple Lie algebras.

Of course, Jacobson's definitive treatise [11] has appeared in the meantime, as have other accounts, including the beginning of Bourbaki's presentation [5]. I feel, nevertheless, that the subject is so important

x

that readers may find still another exposition useful.

I am very grateful to Robert Kibler for reading Chapter I carefully and calling numerous slips to my attention.

The affirmative solution of Hilbert's fifth problem was achieved in 1952 by the combined efforts of Gleason [8] and Montgomery and Zippin [16]. In 1953, two important papers by Yamabe [27],[28] brought significant simplifications, and virtually completed our knowledge of the structure of locally compact groups.

There have been four subsequent published accounts: by Montgomery and Zippin [17, Chapters III and IV], by Gluskov [9], by Shoenfield [25], and by Jacoby [10] (Jacoby's paper extends the theory to local groups).

I gave expositions of Hilbert's fifth problem four times. The first occasion was a course at Chicago in the Spring of 1956. Lars Hörmander was an auditor, and I owe him a great debt for many keen suggestions. (After writing the present account, I learned that Hörmander had also written notes on the subject, and he kindly sent me a copy.) Brief versions of the course were presented at Wisconsin (Summer, 1956) and Princeton (Autumn, 1956). My final effort was tne Winter, 1958 course. Among other things, it drew upon various unpublished notes of Yamabe. But above all it was the fruit of long hours of conversation with Yamabe and with Terry Mirkil. I regret deeply that these notes are appearing years after their untimely deaths.

The reconstruction of the 1958 course after so long a time would have been very difficult without the aid of a superb set of notes taken by Arunas Liulevicius (then a student, now my colleague in the Chicago

Mathematics Department). Hearty thanks go to him for preserving the notes for twelve years and then allowing me to use them.

I hope this fifth account of Hilbert's fifth problem will be useful to the mathematical community. However, I have not at all attempted to make it definitive.

A final remark concerns the style of exposition in Chapter II. Since there are long chains of arguments, numerous lemmas have been inserted. These are stated in naked form, and the reader will have to scan the surrounding context to discover what they say. All theorems, however, are stated in full. (Note: the numbering of theorems and equations begins anew in Chapter II.)

NOTES (ADDED 1974)

Page 85. The work of Kibler mentioned on line 1 has now been published in part in my paper "Infinite-dimensional Lie algebras", Scripta Math. 29(1973), 237-241.

Page 85. In Ex.6 we should assume $p > 3$, since V-algebras were defined only in that case.

Page 88. A good reference for Exs. 4 and 6 is the paper of L. Wallen, "On the magnitude of $x^n - 1$ in a normed algebra", Proc. Amer. Math. Soc. 18(1967), 956. (I am indebted to R. B. Burckel for this reference, and for the keen comments below concerning pages 95, 120, and 129.)

Page 89. The question raised in the first paragraph has been answered in the negative by Sidney A. Morris, "Quotient groups of topological groups with no small subgroups", Proc. Amer. Math. Soc. 31(1972), 625-626. He uses the technique of free topological groups.

The following simple example is due to Jan Hrabowski. Take G to be the additive group of all bounded sequences of real numbers, topologized by the sup norm, and take H to be the subgroup consisting of the elements whose n-th coordinate for every n is a rational number with denominator n. Then G is NSS, H is a closed subgroup of G, but G/H has elements of order 2 arbitrarily close to 0.

Page 95, line -3. The choice $a_i = X_i(1/i)$ does not assure $a_i \to 1$. Instead one should pick m_i going to infinity sufficiently rapidly, and set $a_i = X_i(1/m_i)$.

Page 120, line -9. Since we do not know that m_i approaches ∞ monotonically, the statement that $\{\phi_{h_i}\}$ is a subsequence of $\{\phi_i\}$ is inaccurate. However, it is harmless to take still another subsequence so as to assure monotonicity of m_i.

Page 129, lines 12-14. Lemma 15 should not be cited to justify $N(a) < \infty$; this holds simply because U contains no subgroups. On the other hand, we do need Lemma 15 two lines later to justify $N_i < \infty$.

Irving Kaplansky

Chicago, Illinois

CHAPTER I. LIE ALGEBRAS

1. Definitions and examples

Our beginning point is the concept of a ring: a set with an addition and multiplication satisfying the usual axioms, except that there is no assumption of associativity or of a substitute for associativity.

A <u>Lie ring</u> is a ring L satisfying the following two axioms (for all $a, b, c \in L$):

   (1) $a^2 = 0$ (anti-commutativity),

   (2) $ab \cdot c + bc \cdot a + ca \cdot b = 0$ (Jacobi identity).

If in (1) we replace a by $a + b$ and then delete a^2 and b^2 , we obtain

   (1') $ab = -ba$.

If, in L , $2a = 0$ implies $a = 0$, we can return from (1') to (1).

An <u>algebra</u> over a field F is a ring A which is simultaneously a vector space over F in such a way as to satisfy

$$\lambda \cdot ab = \lambda a \cdot b = a \cdot \lambda b$$

for all λ in F and a, b in A. A Lie ring which is simultaneously an algebra is called a <u>Lie algebra.</u>

There is as yet not much of a coherent theory of general Lie rings or infinite-dimensional Lie algebras. In §§1-6 we shall for the most part be discussing finite-dimensional Lie algebras, but of course there is no point in making this assumption when it is not needed. In several sections there will be a blanket assumption of finite-dimensionality.

Examples. (a) The basic example of a Lie ring is obtained as follows. Let A be any associative ring and introduce in A a new multiplication [ab] = ab - ba; this operation we call commutation. To see that A, thus re-equipped, is a Lie ring requires the verification of

(1") [aa] = 0,

(2") [[ab]c] + [[bc]a] + [[ca]b] = 0.

Of course, (1") is obvious. The (mechanical) verification of (2") is a computation which (to quote Zassenhaus) every mathematician should do once in his life.

The importance of this example is so immense that in the literature on Lie algebras it is customary to use brackets for the operation, even when the object under discussion is an abstract Lie ring and no associative ring is in sight. We are departing from this tradition for two reasons: to shorten the writing, and to emphasize all possible analogies with rings other than Lie rings. The use of brackets will be reserved for the situation where an associative ring is given and the bracket denotes actual commutation.

(b) An important addendum to the previous example is the possibility of taking a Lie subring of A, i.e. an additive subgroup of A closed under commutation. A Lie subring of A need not arise from an associative subring of A, so we have genuinely enlarged the class of examples.

A theorem of Poincaré, Birkhoff, and Witt asserts that any Lie algebra can be obtained in this way (i.e., by exhibiting a suitable associative algebra, changing the operation to commutation, and pass-ing to a suitable Lie subalgebra). A more difficult theorem (due to Ado for characteristic 0 and Iwasawa for characteristic p) states that a finite-dimensional Lie algebra can be obtained in this way from a finite-dimensional associative algebra. These theorems, in principle, enable us to reduce Lie questions to associative ones; but this seldom seems to be useful in proving theorems on Lie algebras.

(c) We mention two especially important examples of Lie rings obtained as Lie subrings of associative rings.

(i) Let A be the algebra of all n by n matrices over a field F, and let L denote the subset of all matrices of trace 0. L is clearly a Lie algebra under commutation.

(ii) Let A be any associative ring admitting an involution * (i.e. an anti-automorphism whose square is the identity), and let L be the set of all skew elements, i.e. elements x satisfying $x^* = -x$. It is readily checked that L is a Lie ring under commutation. An example is the case where A is the algebra of all n by n matrices over a field and * is transposition; L is then the Lie algebra of all n by n skew-symmetric matrices.

(d) Let F be any field and let L be a three-dimensional vector space over F. For multiplication take the standard vector product on L. In terms of a basis a, b, c, we have ab = c, bc = a, ca = b. Then L is a Lie algebra. Anti-commutativity is clear. In verifying the Jacobi identity we note that by multilinearity it is enough to check it on basis vectors. If a basis element is repeated, the Jacobi identity is a consequence of anti-commutativity (see Ex. 2). Thus we need only test the identity on the three different basis vectors. It checks -- in fact every separate term is 0.

(e) A derivation of a ring A is an additive mapping D of A into itself satisfying $(ab)D = aD \cdot b + a \cdot bD$ for all $a, b \in A$. The product of two derivations is useless in the present context. But the commutator of two derivations turns out to be a derivation. Hence the derivations of any ring form a Lie ring.

Suppose that A is associative. For fixed x the mapping $a \to ax - xa$ turns out to be a derivation; we call these derivations $\underline{inner}$. Suppose that A is Lie. Then $a \to ax$ is a derivation, called $\underline{inner}$.

(f) Let F be a field of characteristic p. Form the p-dimensional (associative, commutative) algebra A generated by an element x subject to $x^p = 1$. A derivation of A is completely determined by what it does to x, and its value at x is arbitrary. Write D_i for the derivation sending x into x^{i+1} $(i = 0, \ldots, p-1)$. One finds

$$[D_i D_j] = (i-j)D_{i+j} \qquad (i, j = 0, \ldots, p-1),$$

where the subscript $i+j$ is to be taken mod p.

It turns out that this p-dimensional Lie algebra is simple for $p > 2$. It is called the $\underline{Witt\ algebra}$ and is the starting point for a whole array of simple Lie algebras of characteristic p that have no counterpart for characteristic 0.

(g) A Lie group is a topological group with a neighborhood of the identity homeomorphic to Euclidean space in such a way that the group operations are analytic. Chapter II of these notes is devoted to the proof that the assumption of analyticity is redundant. In any event, there is attached to any Lie group a certain finite-dimensional Lie algebra over the real numbers, and this Lie algebra largely determines the structure of the group (see §11 of Chapter II). This connection between Lie groups and Lie algebras is the main motivation for studying Lie algebras.

Chevalley [6] has attached a Lie algebra to any algebraic matrix group. For characteristic 0 the connection works as effectively as it does for Lie groups, but for characteristic p it is tenuous.

(h) Our final example of a Lie ring is any additive abelian group with all products defined to be 0. Such a Lie ring we shall call abelian.

We conclude this introductory section with a brief survey of Lie algebras of dimension ≤ 3.

Dimension one. There is only one -- the abelian algebra.

Dimension two. If the algebra is not abelian, its square must be one-dimensional. (The square of any algebra is the set of all sums of products.) Pick a basis x, y with x spanning the square; we can normalize so that $xy = x$. The rest of the multiplication table of course reads $x^2 = y^2 = 0$, $yx = -x$. This, then, is the only non-abelian two-dimensional Lie algebra. It admits a concrete representation as the set of all 2 by 2 matrices with second row zero, under commutation.

It is worth noting that up to this point the Jacobi identity has not entered; only anti-commutativity has been used.

Dimension three. We classify according to the dimension of the square L^2 of the given three-dimensional Lie algebra L.

(1) $L^2 = 0$. This, of course, is the abelian case.

(2) L^2 one-dimensional. It is useful to bring the center of L into the discussion, the center of L being defined as the set of all x with

$xL = 0$. If L^2 is non-central, then L is the direct sum of the abelian one-dimensional algebra and the non-abelian two-dimensional algebra. If L^2 is central, say spanned by x, we can complete a basis with elements y and z satisfying $yz = x$. This algebra can be exhibited as all 3 by 3 strictly triangular matrices, i.e. matrices of the form

$$\begin{pmatrix} 0 & a & b \\ 0 & 0 & c \\ 0 & 0 & 0 \end{pmatrix} .$$

We leave the proofs to the reader, and Exs. 4 and 5 offer generalizations.

(3) L^2 two-dimensional. A priori, there are two possibilities: L^2 might be the abelian or the non-abelian two-dimensional algebra. Actually the possibility that L^2 is non-abelian can be ruled out. This is easily checked directly, and a generalization is given in Exs. 6 and 7. Thus L^2 is abelian.

Now take any element z not in L^2. The crucial thing we need to know is how z acts on L^2. Multiplication by z (say on the right) induces a linear transformation on L^2 which must be non-singular, for this is our only chance to get back all of L^2. If we change z by an element of L^2, there is no change in the linear transformation induced on L^2. The remaining change that can be made is to multiply z by a non-zero scalar. So the upshot is that L is determined by a non-singular linear transformation which is unique up to multiplication by a non-zero scalar. This invariant can be alternatively described as an element in the 2 by 2 projective group. If the underlying field admits no quadratic extensions, L has a basis with a fairly simple multiplication table (Ex. 8). Note the appearance of a parameter α, free to range over F. At this point in the classification one begins to encounter families of algebras.

(4) L^2 three-dimensional, i.e. $L^2 = L$. A discussion of the structure of L is best delayed until we have more techniques (see Ex. 16 in § 2). However, we shall state some of the main facts at this point.

I. If every element in F is a square, then L is uniquely determined. It can be uniformly described, for any F, as the vector product algebra (example (d) above). If F has characteristic $\neq 2$, an alternate description is the algebra of 2 by 2 matrices of trace 0.

II. If F is real closed, there are precisely two answers. They can be described as the vector product algebra, and the algebra of 2 by 2 matrices of trace 0.

III. Over other fields the answer may be quite complex, and over bad fields no satisfactory answer has been given at all. Over the field of rational numbers, there are an infinite number of possibilities; they are described by rather subtle invariants. For characteristic $\neq 2$, the problem is identical with the classification of three-dimensional quadratic forms, or of (associative) quaternion algebras.

Exercises

1. Show that the center of an n-dimensional Lie algebra cannot have dimension n-1. More generally, prove that in any (possibly infinite-dimensional) Lie algebra the center cannot be a subspace of codimension 1.

2. Let A be an anti-commutative algebra. Let a, b, c be linearly dependent elements of A. Prove that the Jacobi identity holds for a, b, c.

3. Let L be the algebra of all 2 by 2 matrices of trace 0 over a field of characteristic two. Prove that the center of L is one-dimensional and is equal to L^2.

4. Let L be a Lie algebra whose square is one-dimensional and non-central. Show that L is the direct sum of an abelian algebra and the two-dimensional non-abelian algebra.

5. Let L be a Lie algebra whose square is one-dimensional and central. Then L is the direct sum of an abelian algebra and an algebra with basis x, y_i, z_i where $y_i z_i = x$ and all other products are 0.

6. Prove that all derivations of the two-dimensional non-abelian Lie algebra are inner.

7. Let L be a Lie algebra with the following properties: it is centerless, not equal to its square, and all derivations are inner. Prove that L cannot be the square of a Lie algebra.

8. Let F be a field in which every quadratic equation has a root. Let L be a three-dimensional Lie algebra over F, with L^2 two-dimensional. Prove that L has a basis of one of the following forms:

$$xy = 0, \quad xz = x, \quad yz = \alpha y \quad (\alpha \neq 0 \text{ in } F)$$
$$xy = 0, \quad xz = x + y, \quad yz = y,$$

and that no two of these algebras are isomorphic.

9. Let F be a field of characteristic $\neq 2$. Show that the following statements are equivalent:

(a) The vector product algebra and the algebra of 2 by 2 matrices of trace 0 are isomorphic over F,

(b) -1 is a sum of two squares in F.

10. Let L be a Lie ring and S a subring of L. The <u>normalizer</u> of S is defined to be the set of all x in L with $xS \subseteq S$. Prove that the normalizer of S is a subring. (Hint: use the Jacobi identity.)

2. <u>Solvable and nilpotent algebras</u>

We give the definitions of <u>ideal</u> and <u>simple</u> in the context of general rings. We say that I is a <u>left ideal</u> in a ring A if I is an additive subgroup of A and $ax \in I$ for any $x \in I$ and $a \in A$. <u>Right ideals</u> are defined similarly, and we say that I is a <u>two-sided ideal</u> if it is both a left ideal and a right ideal. In a commutative or anti-commutative ring there is no distinction between the three kinds of ideals.

Standard facts about homomorphisms and quotient rings are valid. If I is a two-sided ideal in a ring A, there is a natural ring structure on A/I and a natural homomorphism from A to A/I. Clearly A/I inherits all identities enjoyed by A; thus if A is a Lie ring, so is A/I. The kernel of a homomorphism on A is a two-sided ideal J, and A/J is isomorphic to the image. If I and J are two-sided ideals, so are $I + J$ and $I \cap J$, and $I/(I \cap J)$ is isomorphic to $(I+J)/J$.

If A is an algebra, then one requires of an ideal that it also be a vector subspace. Scalars float harmlessly through considerations such as those in the preceding paragraph. For definiteness, we shall deal with algebras in the present section. The concept of rings with operators could serve as a unifying device, but for our purposes the extra effort is not worthwhile.

A ring A is <u>simple</u> if there are no two-sided ideals other than 0 and A and furthermore $A^2 = A$. If A is an algebra, then only algebra ideals are being excluded, and it is conceivable that this is a weaker concept; however (Ex. 17) actually there is no difference.

In any algebra A we can form the square A^2 (the set of all sums of products). It is a two-sided ideal with the property that the quotient A/A^2 is trivial (has all products 0); moreover it is the smallest such ideal.

By iterating this construction we form the derived series of A, defined by $A^{(0)} = A$, $A^{(1)} = A^2, \ldots, A^{(n+1)} = (A^{(n)})^2, \ldots$. After $A^{(1)}$, the terms of the derived series need not be two-sided ideals in A (but this is true, by Theorem 4, in a Lie algebra). Each $A^{(n+1)}$ is of course a two-sided ideal in the preceding term $A^{(n)}$.

Definition. An algebra A is solvable if $A^{(n)} = 0$ for some n. The smallest such n is called the length of A.

Remark. There is also a standard concept of solvability for groups, and the use of the same word is of course not a coincidence. Indeed, a connected Lie group is solvable if and only if its Lie algebra is solvable.

Theorem 1. Any subalgebra or homomorphic image of a solvable algebra is solvable. If, for a two-sided ideal I in an algebra A, both I and A/I are solvable, so is A/I.

Proof. If A is solvable of length n, obviously any subalgebra or homomorphic image is solvable with length $\leq n$.

Assume that $I^{(r)} = 0$ and $B^{(s)} = 0$ where $B = A/I$. Then $A^{(s)}$ lies in the kernel of the homomorphism $A \to A/I$, and so $A^{(s)} \subset I$. Hence applying the squaring operation r times to $A^{(s)}$ must yield 0.

This means that $A^{(r+s)} = 0$, so that A is solvable with length at most r + s.

Theorem 1 makes it possible to apply a general type of argument to get Theorem 2.

Theorem 2. If I and J are solvable two-sided ideals in an algebra, then I + J is solvable.

Proof. We have that J is a solvable two-sided ideal in I + J, and that $(I+J)/J$ is solvable, since it is isomorphic to $I/(I \cap J)$, a homomorphic image of I. Hence I + J is solvable.

Theorems 1 and 2 have Theorem 3 as an immediate corollary.

Theorem 3. Any finite-dimensional algebra A has a unique largest solvable two-sided ideal R (that is, R is solvable and contains every solvable two-sided ideal). The algebra A/R contains no nonzero solvable ideals.

When the algebra under discussion is a finite-dimensional Lie algebra L, we call the maximal solvable ideal of L the radical of L, and we say that L is semi-simple if its radical is 0.

The derived series of a Lie algebra consists of ideals. This follows from Theorem 4.

Theorem 4. The product of two ideals in a Lie algebra is again an ideal.

Proof. Let I, J be ideals in the Lie algebra L. Let x ϵ I, y ϵ J, a ϵ L. We have $xy \cdot a + ya \cdot x + ax \cdot y = 0$. Since ya $\cdot$ x ϵ JI = IJ, and ax $\cdot$ y ϵ IJ, we have xy $\cdot$ a ϵ IJ.

Remark. Here it is vital that IJ denotes not just the set of all products ij, but all <u>sums</u> of such products.

The next result is easy but useful.

Theorem 5. The following two statements are equivalent for a Lie algebra L: (a) L has no non-zero solvable ideals, (b) L has no non-zero abelian ideals.

Proof. Of course (a) implies (b). Conversely, suppose that (b) holds and that I is a solvable ideal in L. Suppose that $I^{(n)} = 0$, but $I^{(n-1)} \neq 0$. By Theorem 4, $I^{(n-1)}$ is an ideal in L, and it is abelian.

Because of Theorem 5, to check semi-simplicity of a finite-dimensional Lie algebra it suffices to verify the absence of non-zero abelian ideals.

In general algebras there are, in addition to solvability, the competing notions of <u>nil</u> and <u>nilpotent</u> to consider. An algebra is nil if every element is nilpotent (if power-associativity is not assumed, a meaning must be assigned to this). But the concept is vacuous for Lie algebras: every Lie algebra is nil since the square of every element is 0.

A reasonable meaning to attach to nilpotence of an algebra A is that for some n the product of n elements of A is 0, however associated. But for Lie algebras it is convenient to assume something weaker and then prove that the stronger statement is a consequence.

We define the <u>descending central</u> series of a Lie algebra L inductively by $L_1 = L$, $L_2 = L^2, \ldots, L_{n+1} = LL_n, \ldots$. We say that L is <u>nilpotent</u> if the descending central series reaches 0 in a finite number of steps. The smallest n with $L_n = 0$ is called the <u>nilpotent length</u> (and the length previously defined in connection with the derived series might more carefully be called the <u>solvable length</u>).

We note that $L^{(1)} = L_2 = L^2$, $L^{(2)} = L^{(1)}L^{(1)} \subset LL_2 = L_3$, etc. The next theorem is immediate.

<u>Theorem 6.</u> For any Lie algebra L we have $L^{(n)} \subset L_{n+1}$ for all n. If L is nilpotent, it is solvable.

It can happen that a Lie algebra L has an ideal I with both I and L/I nilpotent, and yet L is not nilpotent. For instance this is the case if L is the non-abelian two-dimensional algebra and I is its one-dimensional square. Thus the proof given for Theorem 2 breaks down if "solvable" is replaced by "nilpotent". Nevertheless the result is true (Theorem 9). The key step is Theorem 8, and Theorem 7 is a prelude to Theorem 8. Note that Theorem 8 also serves to identify nilpotence of a Lie algebra with the stronger type of nilpotence discussed above.

<u>Theorem 7.</u> For any Lie algebra L, and for any positive integers i and j, we have $L_i L_j \subset L_{i+j}$.

<u>Proof.</u> We make an induction on j. For $j = 1$ we have $L_i L_1 = L_i L = L_{i+1}$ by definition. For $j > 1$ we have $L_j = LL_{j-1}$ and by the Jacobi identity

(3)
$$L_i \cdot LL_{j-1} \subset L_i L \cdot L_{j-1} + L_i L_{j-1} \cdot L$$
$$= L_{i+1} L_{j-1} + L_i L_{j-1} \cdot L .$$

The first term on the right of (3) is contained in L_{i+j} by our inductive assumption. The second is contained, again by induction, in $L_{i+j-1} L$, and this equals L_{i+j}.

Theorem 8. Let I be an ideal in a Lie algebra L. Suppose that $x \in L$ is a product of n elements of L (in some association), and that r of these elements lie in I. Then $x \in I_r$.

Proof. Let $x = yz$ be the last multiplication that occurs in forming x. Say that p factors of y and q factors of z lie in I, $p + q = r$. By induction on n we may assume that $y \in I_p$ and $z \in I_q$ (we interpret I_o to be L in case p or q is 0). So $x \in I_p I_q$, which is contained in I_r by Theorem 7.

Theorem 9. The union of two nilpotent ideals in a Lie algebra is nilpotent.

Proof. Let I and J be nilpotent ideals in the Lie algebra L. Suppose that $I_m = 0$ and $J_n = 0$. Then $(I + J)_{m+n-1} = 0$. For if x is a product of m+n-1 elements of $I + J$, it is a sum of terms in each of which either m factors in I or n factors in J must occur. We apply Theorem 8.

Theorem 10 is an immediate corollary of Theorem 9.

Theorem 10. In any finite-dimensional Lie algebra there is a unique largest nilpotent ideal.

It should be noted that if N is the maximal nilpotent ideal of L, it may well happen that L/N possesses non-zero nilpotent ideals. For example, in the non-abelian two-dimensional Lie algebra L the maximal nilpotent ideal is L^2 and L/L^2 is of course abelian.

We proceed to a fundamental theorem, called Engel's theorem, concerning the nilpotence of Lie algebras. Classically the theorem, when stated as in Theorem 13, concerns Lie algebras of nilpotent linear transformations. We give a form which is more general in two respects. First, we work with a Lie "set" rather than a Lie algebra (see the definition that follows). This refinement is due to Jacobson, and has a neat application to automorphisms without fixed points which we give as Theorem 16. Second, we allow the surrounding associative algebra to be infinite-dimensional.

Definition. A Lie set in an associative ring is a subset closed under commutation (but not necessarily under addition).

Remarks. Any non-void Lie set contains 0. For any a, the set $\{0, a\}$ is a Lie set.

Theorem 11. Let A be a (possibly infinite-dimensional) associative algebra. Let S be a Lie set in A, such that every element in S is nilpotent, and the subspace L spanned by S is finite-dimensional. Then S is associative-nilpotent, by which we mean that for some fixed integer n the associative product of any n elements of S is 0.

Proof. Note first that L is a Lie subalgebra of A, and that in the conclusion we might just as well have said that L is associative-nilpotent.

There exist associative-nilpotent Lie subsets of S, for instance 0, or we can harmlessly toss in one more element. Among all such pick one, say T, such that the subspace M it spans has largest possible dimension. If $S \subseteq M$, we are finished. So we assume the contrary. Suppose that the product of any r elements of T (or equivalently M) is 0. Then the following is true for any element x in A: the result of commutating x successively with $2r-1$ elements of M is 0. This is true in particular for a selected element y in S but not in M. Let k be the smallest integer such that commutating y with any k successive elements of T lands again in M. Then a suitable commutation of y with $k-1$ elements of T yields an element z with the following properties: $z \in S$, $z \notin M$, $[zM] \subset M$. Now form the Lie subset of A, say U, generated by T and z, and let N be the subspace it spans. Clearly N is spanned by M and z. We shall prove that N is associative-nilpotent. Since this contradicts the assumed maximal dimension of M, the hypothesis $S \not\subset M$ is untenable.

We first prove the following: let the element u be the product of r elements of M and any number of z's (taken in any order). We claim that $u = 0$, and we argue by induction on the number of z's appear appearing in u. If the number of z's is zero, then $u \in M^r = 0$, and this starts the induction. Now suppose that somewhere in u we have the product mz occurring. Since $mz - zm = m_1 \in M$, and since the result of replacing mz by m_1 in u is 0 (the number of z's has been decreased) we can commute m past z. This means that all r factors from M that occur in u can be pushed to the right past all the z's. This shows that $u = 0$.

We now conclude the proof that N is associative-nilpotent. In fact, if $z^i = 0$, we show that $N^{ir} = 0$. The general element of N has the form $\lambda z + m$ (λ a scalar, $m \in M$). Hence it will suffice for us to examine a product of ir elements, each either z or in M. If we use m factors from M, we get 0, as we have just seen. So only $r-1$ factors from M can be tolerated. This leaves just r vacancies in which to insert a power of z, and we dare not use a higher power than z^{i-1}. The total length that can be assembled without getting 0 is therefore

$$r - 1 + r(i - 1) = ri - 1.$$

This concludes the proof of Theorem 11.

When we specialize Theorem 11 to matrices, we can make an assertion about triangular form.

Theorem 12. Let S be a Lie set of linear transformations on a finite-dimensional vector space V, and suppose that every element of S is nilpotent. Then there exists a basis of V relative to which the members of S assume strict triangular form (zeros on and below the diagonal).

Proof. By Theorem 11 there exists an integer n such that the product of any n elements of S is 0. Take n minimal with this property and pick a vector x such that $y = xT_1 T_2 \cdots T_{n-1} \neq 0$ for suitable $T_i \in S$. Then $yS = 0$. We take y to be the last basis vector. If W is the one-dimensional subspace spanned by y, the hypotheses are preserved when S acts on V/W, and the proof is completed by induction.

We wish to restate Theorem 11 in its usual form, referring to the internal structure of a Lie algebra. We seize this opportunity to introduce representations of Lie algebras.

Definition. Let L be a Lie algebra over a field F. A representation of L is a Lie homomorphism $x \to S_x$ from L into linear transformations on a vector space over F. Note that

$S_{xy} = [S_x S_y] = S_x S_y - S_y S_x$. The $\underline{kernel}$ is the set of all x with $S_x = 0$ and is an ideal in L. If the kernel is 0, the representation is $\underline{faithful}$.

We call a representation of L on a vector space V $\underline{irreducible}$ if the only invariant subspaces are 0 and V, and $\underline{completely\ reducible}$ if V is a direct sum of irreducible subspaces.

We write R_x for the mapping on L given by right-multiplication by x. It is immediate from the Jacobi identity that $x \to R_x$ is a representation of L, and we call it the $\underline{regular}$ representation. Note that the kernel of the regular representation is the center of L. (In the context of Lie algebras the customary term is $\underline{adjoint}$ representation, with notation $ad(x)$. Our reason for departure from custom is the desire to stress the analogy with other algebras). We now state the usual Engel theorem, again with Jacobson's refinement.

$\underline{Theorem\ 13.}$ Suppose a Lie algebra L can be spanned by a subset S closed under multiplication with the property that R_x is nilpotent for every $x \in S$. Then L is nilpotent.

$\underline{Proof.}$ The set of all R_x, $x \in S$, is a Lie set of linear transformations on L satisfying the hypotheses of Theorem 11. Hence for some k, $R_{y_1} R_{y_2} \cdots R_{y_k} = 0$ for all $y_i \in L$. (In fact $k = n$ will do.) From this it follows that $L_{k+1} = 0$; L is nilpotent.

The application of Theorem 13 to automorphisms requires some information on the multiplicative behavior of characteristic subspaces relative to an automorphism. An analogous result on derivations will come first, and it will play a basic role in studying the decomposition of a Lie algebra relative to a Cartan subalgebra.

We recall that, for a characteristic root λ of a linear transformation T on a vector space V, the characteristic subspace V_λ consists of all vectors annihilated by some power of $T - \lambda$. V_λ is a subspace invariant under T. If the underlying field is algebraically closed, V is the direct sum of the V_λ's.

Theorem 14.   Let A be a finite-dimensional algebra (not necessarily Lie or associative) over an algebraically closed field. Let D be a derivation of A. Let $A_\alpha, A_\beta, \ldots$ be the characteristic subspaces relative to D. Then: $A_\alpha A_\beta \subset A_{\alpha+\beta}$ (it being understood that $A_{\alpha+\beta} = 0$ if $\alpha + \beta$ is not a characteristic root of D).

Proof.   For any x, y in A we have

(4) $(xy)(D - \alpha - \beta) = x(D - \alpha) \cdot y + x \cdot y(D - \beta)$.

Suppose that $x(D - \alpha)^r = y(D - \beta)^s = 0$. Then iterated application of (4) shows that $(xy)(D - \alpha - \beta)^{r+s-1} = 0$.

Theorem 15.   Let A be a finite-dimensional algebra over an algebraically closed field, T an automorphism of A, and $A_\alpha, A_\beta, \ldots$ the characteristic subspaces of A under T. Then $A_\alpha A_\beta \subset A_{\alpha\beta}$.

Proof.   This time we use the equation

$(xy)(T - \alpha\beta) = x(T - \alpha) \cdot \beta y + xT \cdot y(T - \beta)$,

and the argument concludes exactly as in Theorem 14.

Theorem 16.   Let L be a finite-dimensional Lie algebra admitting an automorphism T of prime order p leaving fixed only the element 0 (in other words, 1 is not a characteristic root of T). Then L is nilpotent.

Proof. It can be assumed that the base field is algebraically closed (extend both L and T to the algebraic closure; the characteristic roots of T do not change). Let $\varepsilon, \eta, \ldots$ denote the characteristic roots of T; they are p-th roots of unity other than 1. For $x \in L_\varepsilon$, $y \in L_\eta$ we have that y is annihilated by some power of R_x, for by Theorem 15, $yR_x^i \in L_{\eta\varepsilon^i}$ and $\eta\varepsilon^i = 1$ for a suitable i. It follows that R_x is nilpotent, since the L_η's span L. Write $S = L_\varepsilon \cup L_\eta \ldots$, the set-theoretic union. Then S satisfies the hypotheses of Theorem 13. Hence L is nilpotent.

If in Theorem 16 we drop the requirement that T is of prime order, it is no longer possible to conclude that L is nilpotent. It is necessary to go to dimension 4 to get an example. Moreover, such an example must be solvable, for Winter [26] has proved the following: if a finite-dimensional Lie algebra L admits an automorphism leaving only 0 fixed, then L is solvable. Kreknin [13] had previously proved this for automorphisms of finite order. For the case of an automorphism of order 4, Ex.7 provides an easy proof of a slightly sharper result.

For the major remaining theorems of this section, characteristic 0 is needed. We head toward Lie's theorem (Theorem 26), with Theorem 23 as an important prelude.

Theorem 17. Let Z be a linear transformation on a finite-dimensional vector space V over a field of characteristic 0. Suppose that $Tr(Z^n) = 0$ for every positive integer n, where Tr denotes the trace. Then Z is nilpotent.

Proof. We can decompose V into a direct sum $V_1 \oplus V_2$ of sub-spaces invariant under Z in such a way that Z is non-singular on V_1 and nilpotent on V_2. (This is an elementary result in matrix theory. It can be proved briefly from scratch, or deduced from the decomposition of V into characteristic subspaces.) Since the trace of a nilpotent linear transformation is 0, the problem reduces to studying Z on V_1. Thus we may assume that Z is non-singular. Now Z satisfies a polynomial equation (with coefficients in the underlying field). Since Z is non-singular, we can cancel a suitable power of Z and arrange that the constant term of the equation

(5)
$$Z^n + \lambda_1 Z^{n-1} + \ldots + \lambda_n = 0$$

is a non-zero scalar. On taking trace in (5) we get a contradiction.

Theorem 18. Let A_i, B_i $(i = 1, \ldots, r)$ be linear transformations on a finite-dimensional vector space over a field of characteristic 0. Write $Z = \Sigma (A_i B_i - B_i A_i)$, and suppose that Z commutes with every A_j. Then Z is nilpotent.

Proof. We have
$$Z^n = Z^{n-1} \Sigma (A_i B_i - B_i A_i) = \Sigma A_i (Z^{n-1} B_i) - \Sigma (Z^{n-1} B_i) A_i .$$

Hence Z^n is a sum of commutators, and $\mathrm{Tr}(Z^n) = 0$. By Theorem 17, Z is nilpotent.

The next theorem is a generalization of Theorem 4. (To get Theorem 4, take S to be the regular representation of L, restricted to an ideal of L.)

Theorem 19. Let S be a representation of a Lie algebra L on a vector space V, and let I be an ideal in L. Then VS(I) is an invariant subspace of V.

Proof. For $v \epsilon$ V, $x \epsilon$ I, and $y \epsilon$ L, we must prove that $vS_x S_y \epsilon$ VS(I). Now $S_{xy} = [S_x S_y]$, and hence

$$vS_x S_y = v[S_x S_y] + vS_y S_x$$
$$= vS_{xy} + (vS_y)S_x ,$$

which lies in VS(I) since $xy \epsilon$ I.

Theorem 20. Let S be a faithful irreducible representation of a Lie algebra L on a finite-dimensional vector space V, and let I be an ideal in L such that S_x is nilpotent for every $x \epsilon$ I. Then $I = 0$.

Proof. By Engel's theorem (Theorem 13) $S(I)^n = 0$ for a suitable integer n. Take n minimal with this property. Then $VS(I)^{n-1} \neq 0$. By iterated use of Theorem 19, $VS(I)^{n-1}$ is an invariant subspace of V, which must be all of V since V is irreducible. Hence $VS(I) = 0$, and $I = 0$ is a consequence since S is faithful.

Theorem 21. Let L be a Lie algebra with center Z. The following statements are equivalent:

(1) $Z \cap L^2 = 0$ and every abelian ideal of L lies in Z,

(2) L is the direct sum of Z and an algebra with no non-zero solvable ideals.

Remark. We do not use the adjective "semi-simple" to replace the phrase "with no non-zero solvable ideals" since semi-simplicity was officially defined only in the finite-dimensional case.

Proof. (2) implies (1). Suppose that $L = Z \oplus M$ where M has no non-zero solvable ideals. Then $L^2 = M^2$, and hence $Z \cap L^2 = 0$. Any abelian ideal in L projects onto an abelian ideal in M, and hence must lie in the kernel Z.

(1) implies (2). Expand L^2 to a vector space complement M of Z. Since $M \supset L^2$, M is an ideal in L. Thus $L = Z \oplus M$ is a Lie algebra direct sum . By Theorem 5, it remains to prove that M has no non-zero abelian ideals. But any abelian ideal J in M is an abelian ideal in L and therefore contained in Z; hence $J = 0$.

Theorem 22. Suppose that the Lie algebra L possesses a collection $\{J_i\}$ of ideals such that $\bigcap J_i = 0$ and each L/J_i satisfies the conditions in Theorem 21 (either, and hence both). Then L satisfies them too.

Proof. Let Z be the center of L. If $u \in Z \cap L^2$, then u maps into an element lying in both the center and the square of L/J_i. Hence $u \in J_i$, $u = 0$.

Let I be an abelian ideal in L. Then I maps into an abelian ideal of L/J_i. Hence the image of I in L/J_i is central. This means $IL \subset J_i$ for every i, $IL = 0$, $I \subset Z$.

Theorem 23. Let L be a finite-dimensional Lie algebra over a field of characteristic 0. Suppose that L admits a faithful completely reducible representation S on a finite-dimensional vector space V. Then L is the direct sum of its center and a semi-simple algebra.

Proof. We first treat the case where S is irreducible. Let Y be the center of L and take $z \in Y \cap L^2$. Then z can be written

$\Sigma\, a_i b_i$ with $a_i, b_i \in L$. Let Z, A_i, B_i denote the linear transformations representing z, a_i, b_i. Then the hypothesis of Theorem 18 is satisfied and we deduce that Z is nilpotent. Thus the ideal $Y \cap L^2$ is represented by nilpotent matrices. By Theorem 20, $Y \cap L^2 = 0$.

Let I be an abelian ideal in L. The argument of the preceding paragraph applies nearly unchanged to the ideal IL. Hence $IL = 0$ and $I \subset Y$. We have thus verified the two hypotheses of Theorem 21. Theorem 23 stands proved for the case where S is irreducible.

We now allow S to be completely reducible. Thus $V = V_1 \oplus \dots \oplus V_r$ is a direct sum of invariant subspaces V_i on each of which the restriction of S is irreducible. Let J_i denote the kernel of S when S is restricted to V_i. Then $\bigcap J_i = 0$. Moreover L/J_i admits a faithful irreducible representation. Hence the statements of Theorem 21 are valid for each L/J_i. By Theorem 22, the same is true for L. This completes the proof of Theorem 23.

Remark. The direct summand complementary to the center in Theorems 21 and 23 is in fact L^2 (and is therefore unique). This will be known to us as soon as we prove (in the next section) that any semi-simple Lie algebra of characteristic 0 is equal to its square.

Theorem 24 is an immediate corollary of Theorem 23.

Theorem 24. Let L be a solvable Lie algebra over a field of characteristic 0. Assume that L admits a faithful completely reducible finite-dimensional representation. Then L is abelian.

If the base field is algebraically closed we can pass to a more explicit result, known as Lie's Theorem. A very elementary result from linear algebra is first needed.

Theorem 25. Let V be a finite-dimensional vector space over an algebraically closed field. Suppose that V is irreducible under a commutative set S of linear transformations. Then V is one-dimensional.

Proof. If every element of S is a scalar, the result is obvious. Suppose then that $A \in S$ is not a scalar, let c be a characteristic root of A, and set W = the set of all $v \in V$ with $vA = cv$. Then W is a proper subspace of V. The commutativity of S implies that W is invariant under S, a contradiction.

Theorem 26 (Lie): Let L be a solvable Lie algebra of linear transformations on a finite-dimensional vector space V over an algebraically closed field of characteristic 0. Then relative to a suitable basis of V, L can be put in simultaneous triangular form.

Proof. Let W be a minimal invariant subspace of V. Then L restricted to W acts irreducibly on it. It follows from Theorem 24 that L is abelian, modulo its kernel when acting on W. By Theorem 25, W is one-dimensional. This gives us our first basis vector, and the process is continued in obvious fashion.

In the course of the proof of Theorem 23 we noted that the following was a consequence of Theorem 18: if I is an abelian ideal in L, then in any representation of L, IL is sent into nilpotent matrices. We now strengthen this so as to allow I be be solvable.

Theorem 27. Let L be a finite-dimensional Lie algebra over a field of characteristic 0, and let R be the radical of L. Then in any representation of L, LR is represented by nilpotent matrices.

Proof. Let V be the representation space. We first treat the case where V is irreducible. Let J be the kernel of the representation, and write $L_o = L/J$, R_o = the radical of L_o. Then L_o admits a faithful irreducible representation, and hence, by Theorem 23, R_o lies in the center of L_o, $L_o R_o = 0$. Now R maps into R_o in the natural homomorphism from L to L_o. Hence LR ⊂ J, i.e. LR gets represented by 0.

Now we treat the general case and make an induction on the dimension of V. Let W be a proper invariant subspace of V. Then by induction LR is represented by nilpotent linear transformations on W and V/W, and hence also on V itself.

Theorem 28. Let L be a finite-dimensional Lie algebra over a field of characteristic 0, R its radical. Then LR is a nilpotent ideal in L.

Proof. Apply Theorem 27 to the regular representation of L. Since the kernel of the regular representation is the center of L, Ex. 1 finishes the proof.

The next theorem could be dismissed by putting L into triangular form by Lie's theorem. However, this requires an algebraically closed field, and we prefer not to overuse the technique of extending the base field. We therefore give a brief proof that stays within the given field.

Theorem 29. Let L be a solvable Lie algebra of linear transformations on a finite-dimensional vector space V over a field of characteristic 0. Suppose A, B ∈ L and A is nilpotent. Then AB is nilpotent.

Proof. If V is irreducible then L is abelian (Theorem 24) and the result is obvious. So let W be a proper invariant subspace. By induction we have that AB is nilpotent on both W and V/W and hence nilpotent on V.

Let S be a representation of a Lie algebra L. We introduce the form f determined by S: for $x, y \in L$ we set $f(x, y) = Tr(S_x S_y)$. Obviously f is bilinear and symmetric. We check that f is _invariant_: $f(xy, z) = f(x, yz)$. This requires the verification that $(S_x S_y - S_y S_x)S_z$ and $S_x(S_y S_z - S_z S_y)$ have the same trace, which is clear since $S_y(S_x S_z)$ and $(S_x S_z)S_y$ have the same trace.

More generally, we call a symmetric bilinear form f on any algebra A _invariant_ if $f(ab, c) = f(a, bc)$ for all $a, b, c \in A$. In particular, we do this for a Lie algebra whether or not the form comes from a representation.

Theorem 30. Let L be a finite-dimensional Lie algebra over a field of characteristic 0 and R its radical. Let S be a finite-dimensional representation of L with associated form f. Then $f(L^2, R) = 0$.

Proof. For $a, b \in L$ and $x \in R$ we must prove $f(ab, x) = 0$, or alternatively $f(a, bx) = 0$. Form the subspace R_1 spanned by R and a. It is immediate that R_1 is a subalgebra of L. Since $R_1^2 \subset R$, R_1 is solvable. Note that $bx \in R \subset R_1$ and that by Theorem 27, S_{bx} is nilpotent. By Theorem 29, applied to the image of R_1, we deduce that $S_a S_{bx}$ is nilpotent and hence has trace 0.

We conclude the section with an analogue of Theorem 23.

Theorem 31. Let L be a finite-dimensional Lie algebra over a field of characteristic 0. Suppose that L admits a finite-dimensional representation S for which the derived form f is non-singular. Then L is the direct sum of its center and a semi-simple algebra.

Proof. Let R be the radical and Z the center of L. By Theorem 30, $f(L^2, R) = f(L, LR) = 0$. Hence $LR = 0$, $R \subset Z$.

Next let $z \in Z \cap L^2$. By Theorem 18, S_z is nilpotent. For any $a \in L$, S_a and S_z commute, whence $S_z S_a$ is nilpotent. Hence $f(z, L) = 0$, $z = 0$. We have now verified the hypotheses of Theorem 21.

Exercises

1. Let I be a central ideal in a Lie algebra L. If L/I is nilpotent, prove that L is nilpotent.

2. In a Lie algebra: (a) prove that the annihilator of an ideal is an ideal, (b) prove that the annihilator of a subalgebra is a subalgebra, (c) give an example where the annihilator of a subalgebra is not an ideal.

3. Let A be a Lie ring or an associative ring. Prove that the inner derivations of A form an ideal in the Lie ring of all derivations of A.

4. Prove that over any field there are exactly two three-dimensional nilpotent Lie algebras.

5. Prove that any four-dimensional nilpotent Lie algebra has a three-dimensional abelian ideal. Use this to classify four-dimensional nilpotent Lie algebras.

6. If a finite-dimensional Lie algebra L of characteristic 0 admits a non-singular derivation D, prove that L is nilpotent. (Split L into D-characteristic subspaces and argue as in Theorem 16.)

7. Let L be a finite-dimensional Lie algebra admitting an automorphism T which leaves only 0 fixed and satisfies $T^4 = I$. Prove that L^2 is nilpotent. (Let the characteristic subspaces under T be L_i, L_{-i} and L_{-1}. Note that $L_i L_{-i}$ and L_{-1}^2 are 0. Argue (by the Jacobi identity) that L_i^2 annihilates L_{-i}, L_{-1}, L^2. Use Theorem 11 to see that L^2 is nilpotent.)

8. Let L be a finite-dimensional Lie algebra over an algebraically closed field of characteristic 0. Suppose that L has a solvable subalgebra S of dimension n, $S \neq L$. Prove that L has a subalgebra of dimension $n+1$. (Use the natural representation of S on L/S.)

9. Let L be an n-dimensional Lie algebra over an algebraically closed field of characteristic 0. Let M be a maximal subalgebra of L. If M is solvable, prove that the dimension of M is $n-1$. (Use Ex. 8.)

10. Let L be a non-nilpotent finite-dimensional Lie algebra over an algebraically closed field. Prove that L has a two-dimensional non-abelian subalgebra. (Pick R_x non-nilpotent and y a characteristic vector for a non-zero characteristic root of x.)

11. Let L be a non-abelian finite-dimensional Lie algebra over an algebraically closed field. Prove that for some $y \in L$, R_y has a non-linear elementary divisor for the characteristic root 0. (If L is non-nilpotent, use the y of Ex. 10.)

12. Let S be a finite-dimensional representation of a Lie algebra, and let f be the attached invariant form. If f is non-singular, prove

that S is faithful.

13. Let f be an invariant symmetric bilinear form on an algebra A. Let I be a left-ideal in A and J the orthogonal complement of I (the set of all x satisfying $f(x, I) = 0$). Prove that J is a right-ideal in A.

14. Let A be any algebra (not necessarily finite-dimensional; there is no assumption that A is Lie or associative). The <u>total annihilator</u> Z of A is the set of x in A with $xA = Ax = 0$. Prove that A has a decomposition $A = B \oplus C$ where B is trivial (all products 0) and the total annihilator of C is contained in C^2. (Take B to be a complement of $Z \cap A^2$ within Z; take C satisfying $C + (Z + A^2) = A$, $C \cap (Z + A^2) = A^2$.)

15. The <u>centroid</u> of an algebra A is the set of all linear transformations on A that commute with all left and right multiplications. Note that the centroid is an algebra containing the identity linear transformation. If A has a unit element, the centroid coincides with the elements that commute and associate with everything. We place centroid elements on the right, denoting a typical one by θ.

(a) Let f be an invariant symmetric bilinear form on A, and suppose that $A^2 = A$. Define g by $g(a, b) = f(a\theta, b)$, where θ lies in the centroid of A. Prove that g is symmetric and invariant.

(b) Suppose that f is a non-singular invariant symmetric bilinear form on A. Assume that A is finite-dimensional. Prove that any invariant symmetric bilinear form g is given by $g(a, b) = f(a\theta, b)$ where θ is a unique centroid element. (Fix an element a. The mapping $b \to g(a, b)$ is representable by taking the inner product under f with a unique element a'. The mapping $a \rightarrow a'$ is a linear transformation which we call θ. That θ lies in the centroid is a routine verification.)

16. Let L be a three-dimensional Lie algebra over a field F. Assume that $L = L^2$.

(a) Suppose that every element of F is a square and that F has characteristic $\neq 2$. Prove that L is uniquely determined as (for instance) all 2 by 2 matrices of trace 0. (Hint: by Engel's theorem there is an element h with R_h non-nilpotent. $\text{Tr}(R_h) = 0$, so we can arrange that the characteristic roots of R_h are $0, 1, -1$. Pick elements x, y with $xh = x$, $yh = -y$. By Theorem 14, xy is a scalar multiple of h. The scalar is non-zero, for otherwise L^2 would fall short of L. We can change x so that $xy = h$.)

(b) Assume again that every element of F is a square and that F has characteristic 2. Prove that L is uniquely determined as (for instance) the vector product algebra of §1. (Hint: again choose h with R_h non-nilpotent and with characteristic roots $0, 1, 1$. If h has simple elementary divisors, we complete a basis with x, y satisfying $xh = x$, $yh = y$. As in part (a), xy can be normalized to be h. Set $k = h + x + y$. Computation shows that R_k has characteristic roots $0, 1, 1$ with a non-simple elementary divisor. A basis can be completed with u, v so that $uk = v$, $vk = u$, $uv = k$.)

(c) Assume that F is real closed. Prove that L is isomorphic either to 2 by 2 matrices of trace 0 or the vector product algebra. (Hint: pick R_h non-nilpotent. If the characteristic roots are in the field, proceed as in (a). Otherwise (after changing h) we reach $xh = y$, $yh = -x$. The element xy is a non-zero scalar multiple of h which can be normalized to ±h, corresponding to the two cases.)

17. Let A be an algebra, and suppose that A is simple in the sense of having no non-trivial ideals that are subspaces. Prove that A is simple as a ring. (Hint: if I is a two-sided ring ideal different from 0, let J be the subspace spanned by I. Use J = A and A = A^2 to deduce that I = A.)

3. Semi-simple algebras

The theorem with which we begin this section (Theorem 32) will doubtless look strange at a first glance. The motivation for it will not become apparent until we are deep in the proof of Theorem 33. However, orderly exposition dictates that Theorem 32 be proved first.

Theorem 32. Let L be an n-dimensional Lie algebra over an algebraically closed field of characteristic 0. Suppose that L has an (n-1)-dimensional solvable subalgebra M and that L is spanned by M and y. Let x be an element of M $\cap$ L^2 such that yx - y ϵ M. Let S be an irreducible representation of L on a vector space V of dimension at least two, and let f be the associated form. Then: f(x, x) is a positive rational number.

Proof. When the representation is restricted to M there is, by Theorem 26, a joint characteristic vector u. Write u_0 = u, u_1 = uS_y, ..., u_{i+1} = u_iS_y, ... We claim that the subspace spanned by u_0, ..., u_i is invariant under S(M). We take an arbitrary element z in M and have to prove that the subspace in question is invariant under S_z. Now any element in L can be written as the sum of an element in

M and a scalar multiple of y. This applies in particular to yz, and so we have $yz - \lambda y \in M$ for a suitable scalar λ. Suppose further that $uS_z = \beta u$. Then we shall prove more precisely that $u_i S_z = (\beta + i\lambda)u_i + $ a linear combination of $u_o, \ldots, u_{i-1}$. Supposing that this is true for i, we prove it for $i+1$. We have $u_{i+1} S_z = u_i S_y S_z = u_i (S_{yz} + S_z S_y)$. Now $yz = \lambda y + m$, $m \in M$, and $u_i S_m$ is by our inductive assumption a linear combination of $u_o, \ldots, u_i$. Thus $u_i S_{yz}$ contributes λu_{i+1} and $u_i S_z S_y$ contributes $(\beta + i\lambda)u_{i+1}$. Together, we get $[\beta + (i+1)\lambda]u_{i+1}$ modulo the subspace spanned by $u_o, \ldots, u_i$.

Pick the largest k such that $u_o, u_1, \ldots, u_k$ are linearly independent. Then the subspace they span is invariant under both S_y and $S(M)$ and hence under L; by the irreducibility of V we have that the elements $u_o, \ldots, u_k$ form a basis of V. Relative to this basis S_x is in triangular form and, by what we showed above, the characteristic roots of S_x are $\alpha, \alpha+1, \alpha+2, \ldots$ where $uS_x = \alpha u$ (note that the λ of the above discussion is 1 in the case of x). Since $x \in L^2$, we have $Tr(S_x) = 0$, i.e.

$$\alpha + (\alpha + 1) + \ldots + (\alpha + k) = 0.$$

Since V is at least two-dimensional, we have $k \geq 1$, and it follows that α is a non-zero rational number. Finally, $f(x, x) = Tr(S_x^2) = $ the sum of the squares of $\alpha, \alpha+1, \ldots$ Hence $f(x, x)$ is a positive rational number.

Theorem 33 (Cartan): Let L be a Lie algebra of matrices over a field of characteristic 0. Suppose that $Tr(AB) = 0$ for every $A, B \in L$. Then L is solvable.

Proof. We can assume the base field to be algebraically closed (take the space spanned by L over the algebraic closure).

We argue by induction on the dimension of L. If $L^2 \neq L$ then L^2 is solvable and so is L. We therefore assume $L^2 = L$ and ultimately reach a contradiction. By the induction again, every proper subalgebra of L is solvable. Pick a maximal subalgebra M. Then (see Ex. 9 in §2) the dimension of L exceeds the dimension of L exceeds the dimension of M by just one. Pick any Y in L but not in M. It is impossible for M to be an ideal in L; therefore there exists $X \in M$ with $[YX] \notin M$. By multiplying X by a suitable scalar we can arrange $[YX] - Y \in M$.

Let V be the vector space on which L acts and pick a composition series $0 = V_0 \subset V_1 \subset \ldots \subset V_r = V$, i.e. a series of invariant subspaces such that L acts irreducibly on each V_{i+1}/V_i. It cannot be the case that each V_{i+1}/V_i is one-dimensional, for then (since $L = L^2$) L acts trivially on each V_{i+1}/V_i and L is nilpotent. So at least one V_{i+1}/V_i has dimension greater than one. On each of these the contribution to $Tr(X^2)$ is a positive rational number by Theorem 32. On the one-dimensional pieces the contribution to $Tr(X^2)$ is 0. The grand $Tr(X^2)$ is the sum of the separate contributions and is therefore not 0. This is the desired contradiction.

We readily deduce a basic theorem.

Theorem 34. Let L be a semi-simple Lie algebra over a field of characteristic 0, S a faithful representation, and f the associated form. Then f is non-singular.

Proof. The set of all x with $f(x, L) = 0$ is an ideal I in L (Ex. 13 in §2). S(I) is a Lie algebra of matrices to which Theorem 33 applies. Hence I is solvable, $I = 0$.

We apply Theorem 34 to the regular representation. The form attached to the regular representation is called the Killing form. Our notation for it will be (x, y).

Theorem 35. Let L be a finite-dimensional Lie algebra over a field of characteristic 0. Then L is semi-simple if and only if its Killing form is non-singular.

Proof. If L is semi-simple, its Killing form is non-singular by Theorem 34. Conversely, assume that the Killing form of L is non-singular. It follows that L is centerless (compare Ex. 12 in §2). By Theorem 31, L is semi-simple.

Remark. The half of Theorem 35 asserting that non-singularity of the Killing form implies semi-simplicity was accessible right after Theorem 31. See Ex. 3 for an alternate proof that it valid in greater generality.

We proceed to the decomposition of a semi-simple Lie algebra into simple components.

Theorem 36. Let A be a finite-dimensional algebra possessing a non-singular invariant form f, and suppose that A contains no non-zero two-sided ideal I satisfying $I^2 = 0$. (In particular, A can be a semi-simple Lie algebra over a field of characteristic 0). Then:

(1) A is a direct sum of simple algebras, the decomposition being also an orthogonal one relative to f,

(2) $A^2 = A$,

(3) any ideal of A is a direct summand of A.

Proof. Let J be a minimal two-sided ideal in A (that is, $J \neq 0$ and there is no two-sided ideal properly between 0 and J). Let J' denote the orthogonal complement of J relative to f. Then J' is a two-sided ideal in A (compare Ex. 12 in §2). We claim that $J \cap J' = 0$. If not, the minimality of J tells us that $J \subset J'$. This means that $f(J, J) = 0$, whence $f(J^2, A) = f(J, JA) = 0$. Hence $J^2 = 0$, a contradiction. We have sustained the claim that $J \cap J' = 0$, i.e., f is non-singular on J. By the basic theory of inner products, A is the vector space direct sum of J and J' (for this it is unnecessary for f to be non-singular on A and A may be allowed to be infinite-dimensional; what is needed is that J is finite-dimensional and that the form is non-singular on J).

The hypotheses of the theorem are inherited by J', and the process continues till A is a direct sum of simple algebras. Conclusions (2) and (3) are immediate consequences of (1).

We can define the Killing form on an arbitrary algebra by $(x, y) = Tr(R_x R_y)$. It is bilinear and symmetric, but we have to abandon invariance.

Theorem 37. Let A be any algebra, I a left ideal in A. Then the Killing form on I (regarded as an algebra in its own right) coincides with the Killing form on A restricted to I.

Proof. Choose a basis of I and enlarge it to a basis of A. For $x \in I$, R_x has the form

$$\begin{pmatrix} * & & 0 \\ * & & 0 \end{pmatrix}$$

Thus for $x, y \in I$, $\mathrm{Tr}(R_x R_y)$ is the same, whether computed for the entire matrix or just for the upper left corner.

Theorem 38. Let L be a finite-dimensional Lie algebra and I an ideal in L. Assume that the Killing form on I (as an algebra in its own right) is non-singular. Then I is a direct summand of L.

Proof. By Theorem 37 the Killing form on L is non-singular when restricted to I. Just as in the proof of Theorem 36, L is a Lie algebra direct sum of I and its orthogonal complement.

Theorem 39. Let L be a finite-dimensional Lie algebra with a non-singular Killing form. Then any derivation of L is inner.

Proof. Let C be the Lie algebra of derivations of L. The mapping $x \to R_x$ is a homomorphism of L into C with kernel the center of L. Since L is centerless, the mapping is an isomorphism of L into C. The image is the algebra of inner derivations of L, which (Ex. 3 in §2) is an ideal in C. By Theorem 38, L (regarded as embedded in C in this way) is a direct summand of C, say $C = L \oplus M$. The elements of M are derivations which commute with all inner derivations. Thus if $D \in M$ we have $DR_x = R_x D$ for all x in L. Now $R_{xD} = [R_x, D]$; this is a restatement of the property that D is a derivation. Thus we reach $R_{xD} = 0$, $xD = 0$, $D = 0$. We have shown that $M = 0$, and so $C = L$, all derivations are inner.

Exercises

1. Let A be any finite-dimensional algebra, and let I be a two-sided ideal in A satisfying $I^2 = 0$. Prove that $(I, A) = 0$, where $(,)$ denotes the Killing form. (Hint: pick a basis of A by enlarging a basis of I. For $x \epsilon I$, $a \epsilon A$, R_x and R_a have the form

$$\begin{pmatrix} 0 & 0 \\ * & 0 \end{pmatrix} \quad , \quad \begin{pmatrix} * & 0 \\ * & 0 \end{pmatrix}$$

Hence $Tr(R_x R_a) = 0$.)

2. Let A be a finite-dimensional algebra with a non-singular Killing form. Prove that A does not possess a non-zero two-sided ideal I satisfying $I^2 = 0$. (Hint: use Ex. 1).

3. Let F be any field (not necessarily of characteristic 0). Let L be a finite-dimensional Lie algebra over F, and suppose that the Killing form of L is non-singular.

(a) Prove that L is semi-simple. (Hint: use Ex. 2 and Theorem 5).

(b) Prove that the three conclusions of Theorem 36 are valid for L.

4. Let L be a finite-dimensional Lie algebra over a field of characteristic 0, and let R be its radical (maximal solvable ideal). Prove that R is the orthogonal complement of L^2 relative to the Killing form. (Hint: $(L^2, R) = 0$ by Theorem 30. Conversely, let I be the orthogonal complement of L^2. I is an ideal and we have to prove that it is solvable. We have $(IL, L) = 0$. By Theorem 33, the image of IL in the regular representation is solvable. Since the kernel is central, IL is solvable. Hence I^2 is solvable, I is solvable.)

5. Let L be any finite-dimensional Lie algebra and D a derivation of L. Prove that $(aD, b) = -(a, bD)$ for any $a, b \in L$. (Hint: that D is a derivation is equivalent to $R_{xD} = [R_x, D]$. The statement $\text{Tr}(R_{aD}R_b) + \text{Tr}(R_a R_{bD}) = 0$ admits a routine verification.)

6. Let L be a finite-dimensional Lie algebra over a field of characteristic 0. Let R be the radical of L, and let D be a derivation of L. Prove that D carries R into itself. (Hint: use Exs. 4 and 5.)

7. In Ex. 6 let N be the maximal nilpotent ideal of L. Prove that D carries R into N. (Hint: adjoin D to L. Use Ex. 6 and Theorem 28.)

8. Let L be as in Ex. 6 and let I be an ideal in L. Prove that $R \cap I$ is the radical of I. (Hint: deduce from Ex. 6 that the radical of I is an ideal in L.)

4. Cartan subalgebras

We begin by showing that the decomposition of a vector space into characteristic subspaces can be generalized from a single linear transformation to a whole nilpotent Lie algebra of linear transformations. We first prove a preliminary result.

Theorem 40. Let S and T be linear transformations satisfying
$$[\ldots [[ST]T]\ldots] = 0,$$
i.e., the result of commuting S with T sufficiently often is 0. Then any characteristic subspace of T is invariant under S.

Proof. Nothing is changed if we subtract a scalar from T. Consequently we may assume that we are given a vector x annihilated by some power of T, and have to prove that the same is true for xS. In fact, if $xT^r = 0$ we shall prove $xST^{n+r-1} = 0$ where n is the number of copies of T occurring in the hypothesis. To do this we prove by induction on i that $xT^{r-i}ST^{n+i-1} = 0$ for $i \leq r$. For $i = 0$ it is certainly true. We suppose it known up to and including i and prove it for $i+1$. Now it follows from the hypothesis of the theorem that for any vector y, yST^n is a linear combination of terms $yT^a ST^b$ with $a + b = n$, $a \geq 1$. We apply this with $y = xT^{r-i-1}$. We deduce that $xT^{r-i-1}ST^{n+i}$ is a linear combination of terms

$$xT^{r-i-1+a}ST^{i+b} \qquad (a+b = n, \quad a \geq 1).$$

These terms are all 0 by the inductive assumption. Thus $xT^{r-i-1}ST^{n+i} = 0$, which is the statement we needed to prove for $i+1$. When we reach $i = r$ we have the desired conclusion.

By repeated use of Theorem 40 we obtain the next theorem.

Theorem 41. Let L be a nilpotent Lie algebra of linear transformations on a finite-dimensional vector space V over an algebraically closed field. Then V is a direct sum of subspaces invariant under L on each of which every member of L has the form scalar plus nilpotent.

The hypothesis of nilpotence in Theorem 41 cannot be dropped. Indeed a Lie algebra of linear transformations satisfying the conclusion of Theorem 41 is necessarily nilpotent.

The following definition suggests itself.

Definition. Let S be a set of linear transformations on a vector space over a field K. A function λ from S to K is called a <u>root</u> if there exists a non-zero vector which, for every $T \in S$, is annihilated by some power of $T - \lambda(T)$. The set of all such vectors is a subspace called the <u>root space</u> for λ.

In this terminology, Theorem 41 simply says that V is spanned by the root spaces for the roots of L. If we group together all subspaces corresponding to a single root, the decomposition is unique.

A root is automatically homogeneous relative to scalars, but (for characteristic $p \neq 0$) it need not be additive. For example, let K be a perfect field of characteristic 2 ("perfect" means that every element in K is a square in K), and let L be the Lie algebra of all 2 by 2 matrices of trace 0 over K. Then the function assigning to each matrix its unique characteristic root is a root on L but it is not additive. In Theorem 42 we note two cases where we can prove that roots are additive.

Theorem 42. Let L be a Lie algebra of linear transformations on a finite-dimensional vector space over a field K, and let λ be a root on L. If the characteristic of K is 0, or if L is abelian regardless of the characteristic, then λ is linear.

Proof. Let x be the vector exhibiting the fact that λ is a root. By dropping down to the invariant subspace spanned by x, we can assume that for each $T \in L$, $T - \lambda(T)$ is nilpotent. By extending K we can assume K to be algebraically closed (not really necessary since all characteristic roots are in K anyway). The result is now evident from the possibility of simultaneous triangular form, using Theorems 26 and 25 respectively.

Theorem 43. Let E be a nilpotent Lie algebra of derivations of a finite-dimensional algebra A over an algebraically closed field. Let $A_\alpha, A_\beta, \ldots$ be the root spaces for the roots $\alpha, \beta, \ldots$ of E. Then $A_\alpha A_\beta \subset A_{\alpha+\beta}$ (it being understood that the product is 0 if $\alpha+\beta$ is not a root).

Proof. This is immediate from Theorem 14.

Theorem 43 acquires greater interest if A is a Lie algebra, because of the possibility of taking E to be a nilpotent subalgebra of A (inducing inner derivations via right-multiplication). In that case there is a non-zero element of E annihilated by E, and so the function assigning 0 to every element of E is sure to be a root. If we write A_α for the root spaces under E then, by Theorem 43, A_0 is a subalgebra. Of course A_0 contains E. The case where it equals E is so important that it is singled out for a definition.

Definition. Let L be a finite-dimensional Lie algebra over an algebraically closed field. A subalgebra H is a Cartan subalgebra if H is nilpotent and if, when L is decomposed relative to H, the 0-subalgebra is H.

Remarks. 1. A Cartan subalgebra is a maximal nilpotent subalgebra (Ex. 1), but the converse is false (Ex. 2).

2. An alternative characterization of a Cartan subalgebra is that it is a nilpotent subalgebra which is equal to its normalizer (Ex. 3).

It is by no means evident that Cartan subalgebras exist. This is proved by a basic construction due to Cartan.

Let L be a Lie algebra over an algebraically closed field. Call an element u of L <u>regular</u> if R_u has the smallest possible number of 0 characteristic roots; this minimal number is called the <u>rank</u> of L. (Note that the rank is at least one, for $uR_u = 0$ for any u). Perform the decomposition of L relative to R_u, and let H be the subalgebra corresponding to the characteristic root 0. We prove in Theorem 45 that H is a Cartan subalgebra. Theorem 44 is an elementary prelude.

<u>Theorem 44.</u> Let A and B be n by n matrices over any field. Suppose that $kA + B$ is nilpotent for $n+1$ distinct scalars k. Then B is nilpotent.

<u>Proof.</u> On expanding $(kA + B)^n = 0$ we get

$$k^n A^n + k^{n-1}(BA^{n-1} + ABA^{n-2} + \ldots + A^{n-1}B) + \ldots + B^n = 0.$$

This gives us $n+1$ linear homogeneous equations in $n+1$ "unknowns". The matrix of coefficients is a Vandermonde matrix and therefore non-singular. Hence the $n+1$ unknowns vanish, and in particular $B^n = 0$.

<u>Theorem 45.</u> Let L be a finite-dimensional Lie algebra over an algebraically closed field. Let u be a regular element of L. Let H be the subalgebra corresponding to the characteristic root 0 of R_u. Then H is a Cartan subalgebra of L.

<u>Proof.</u> Let

$$L = L_o + L_\alpha + L_\beta + \ldots$$

be the decomposition of L relative to R_u, and note that we are writing H for L_o. In order to prove H nilpotent it suffices, by Engel's theorem (Theorem 13), to prove that all right multiplications by elements of H are nilpotent when they act on H. For any v in H we have that L_α is invariant under R_v (since $L_\alpha L_o \subset L_\alpha$ by Theorem 43). For only a finite number of scalars k is it the case that $kR_u + R_v$ is singular when restricted to L_α. To see this, observe that R_u is non-singular on L_α (it acts on L_α with the single non-zero characteristic root α), and that $kR_u + R_v$ is singular if and only if k is a characteristic root of $-R_u^{-1}R_v$. An infinite number of scalars is available, since the base field is algebraically closed. Thus we can find an infinite number of k's such that $kR_u + R_v$ is non-singular on each L_α with $\alpha \neq 0$. But then such a $kR_u + R_v$ must be nilpotent on H, for otherwise ku + v would have fewer 0 characteristic roots in its right multiplication than R_u. By Theorem 44, R_v is nilpotent on H, as required.

Once it is known that H is nilpotent, it is immediate that H is a Cartan subalgebra, for the 0-subalgebra under H is part of the 0-subalgebra under u and cannot be larger than H. This concludes the proof of Theorem 45.

We now investigate how the decomposition relative to a Cartan subalgebra meshes with an invariant form. In Theorem 46 the form need not come from a representation.

Theorem 46. Let L be a finite-dimensional Lie algebra over an algebraically closed field. Let H be a Cartan subalgebra of L, and let $L_\alpha, L_\beta, \ldots$ be the root spaces relative to H. Let f be an invariant form on L. Then: (a) $f(L_\alpha, L_\beta) = 0$ for $\beta \neq -\alpha$, (b) if f is non-singular, then L_α and $L_{-\alpha}$ are dual spaces relative to f, and f is non-singular on H.

Proof. First we prove that $f(L_\alpha, H) = 0$ for $\alpha \neq 0$. Given $h \in H$ and $x \in L_\alpha$ we have to prove that $f(x, h) = 0$. Pick an element k in H with $\alpha(k) \neq 0$. Then R_k is non-singular on L_α. Hence $x = yR_k$ for a suitable y in L_α, and $f(x, h) = f(yk, h) = f(y, kh)$. Since H is nilpotent, iteration of this procedure sufficiently many times results in $f(x, h) = 0$.

Now let $x \in L_\alpha$, $t \in L_\beta$, and again write $x = yk$ as above. Then $f(x, t) = f(yk, t) = -f(ky, t) = f(k, ty)$. By what we just showed, this is 0 unless $ty \in H$, i.e., $\beta = -\alpha$. We have proved part (a) of the theorem, and part (b) is an immediate consequence.

For the next theorem (a companion result to Theorem 30) we assume that the form comes from a representation.

Theorem 47. Let L be a finite-dimensional Lie algebra over an algebraically closed field, Z its center, S a representation of L, and f the form arising from S. Assume either that the characteristic is 0 or that L is nilpotent. Then $f(Z \cap L^2, L) = 0$.

Proof. We may assume that S is irreducible. Let z be an element of $Z \cap L^2$. By Schur's lemma, $S(z)$ is a scalar matrix. On the other hand z lies in L^2 and therefore $S(z)$ has trace 0. If the characteristic is 0 we conclude that $S(z) = 0$ and the proof is finished. So we assume characteristic $\neq 0$, and have by hypothesis that L is nilpotent. In any event more discussion is needed only for $S(z) \neq 0$, so we have that the dimension of the representation space is a multiple of the characteristic. (Actually it can be shown to be a power of the characteristic, but we do not need this refinement.) By Theorem 41, we have that for any $a \in L$, $S(a)$ has the form scalar plus nilpotent. It follows that

$$f(z, a) = Tr[S(z)S(a)] = 0 ,$$

as required.

For characteristic $\neq 0$ we cannot, in Theorem 47, delete the assumption that L is nilpotent. In fact it does not even suffice for L to be solvable: let L be the Lie algebra of all 2 by 2 matrices over a field of characteristic 2, and take S to be the identity representation. Then for

$$z = \begin{pmatrix} 1 & 0 \\ 0 & 1 \end{pmatrix} , \qquad a = \begin{pmatrix} 1 & 0 \\ 0 & 0 \end{pmatrix}$$

we have $z \in Z \cap L^2$, but $f(z, a) \neq 0$.

In any non-abelian nilpotent Lie algebra L we have $Z \cap L^2 \neq 0$ (take the penultimate term of the descending central series). Hence:

Theorem 48. Let L be a finite-dimensional nilpotent Lie algebra over an algebraically closed field, let S be a representation of L, and let f be the form arising from S. Suppose that f is non-singular. Then L is abelian.

In Theorems 47 and 48 it does not suffice to assume that f is merely an invariant form (i.e., one not necessarily coming from a representation). The lowest dimension for a counterexample is 6. Take as basis z, a, b, c, d, u with $ab = z$, $au = c$, $bu = d$ and all other products 0; take $f(z, u) = f(a, d) = -f(b, c) = 1$ and all other inner products 0. Then L is nilpotent, $z \in Z \cap L^2$, but $f(z, u) \neq 0$. Needless to say, this example provides us with an invariant form that cannot come from any representation.

By combining Theorems 46 and 48 we obtain:

Theorem 49. Let L be a finite-dimensional Lie algebra over an algebraically closed field. Assume that L possesses a representation whose attached form is non-singular (this is true if the Killing form of L is non-singular, which in turn is true if L is semi-simple and the characteristic is 0). Then any Cartan subalgebra of L is abelian.

Exercises

1. Prove that a Cartan subalgebra is maximal among nilpotent sub-algebras. (Hint: a larger nilpotent subalgebra of a Cartan subalgebra H would be part of the 0-subalgebra of H.)

2. Let L be the Lie algebra of 2 by 2 matrices of trace 0 over an algebraically closed field of characteristic $\neq 2$. Prove that the sub-algebra spanned by a nilpotent non-zero matrix is maximal nilpotent but is not Cartan.

3. Let H be a nilpotent subalgebra of a finite-dimensional Lie algebra L over an algebraically closed field. Prove that H is Cartan if and only if H is its own normalizer (the normalizer of H is the set of all a with aH $\subset$ H).

5. Transition to a geometric problem (characteristic 0)

In this section we shall pursue the study of semi-simple Lie alge-bras of characteristic 0 up to the point where we find a related geo-metric structure. For the first theorem we do not invoke semi-simplicity.

Theorem 50. Let L be a finite-dimensional Lie algebra over an algebraically closed field of characteristic 0, let H be a Cartan subalgebra of L, let α be a non-zero root, let β be any root, and let $h = xy$ with $x \in L_\alpha$, $y \in L_{-\alpha}$. Then $\beta(h)$ is a rational multiple of $\alpha(h)$.

Proof. Let V be the subspace of L spanned by all $L_{\beta + i\alpha}$ where i ranges over the integers. It is evident that V is invariant under R_x and R_y. Since $R_h = R_x R_h - R_y R_x$, we have that the trace of R_h restricted to V is 0. Now on any root space L_γ, R_h has the form $\gamma(h)I + $ nilpotent. Hence $\Sigma \gamma(h)Dim(L_\gamma) = 0$, where the sum is taken over all $\gamma = \beta + i\alpha$. Since $\gamma(h) = \beta(h) + i\alpha(h)$, we deduce an equation of the form $r\beta(h) = s\alpha(h)$ with r and s integers. In particular, r is the sum of the dimensions of the spaces $L_{\beta + i\alpha}$ and therefore is a positive integer. Hence $\beta(h)$ is a rational multiple of $\alpha(h)$.

Theorem 51. In the setup of Theorem 50 assume in addition that L is semi-simple and that $\alpha(h) = 0$. Then $h = 0$.

Proof. By Theorem 35 the Killing form on L is non-singular. Hence it suffices to prove $(h, L) = 0$, and by Theorem 46 it further suffices to prove $(h, k) = 0$ for $k \in H$. We look at R_h and R_k on a root space L_β. By Theorem 50 and our hypothesis, $\beta(h) = 0$. Since H is abelian (Theorem 49), R_h and R_k commute. (If preferred, we could instead use Lie's theorem -- Theorem 26 -- here.) We have that R_h is nilpotent on every L_β; hence so is $R_h R_k$. Thus $Tr(R_h R_k) = 0$ on every root space, and therefore $Tr(R_h R_k)$ vanishes on all of L. We have proved that $(h, k) = 0$, as required.

Theorem 52. Let L be a semi-simple Lie algebra over an alge-braically closed field of characteristic 0, and let H be a Cartan sub-algebra of L. Then the root spaces for non-zero roots relative to H are one-dimensional. No integral multiple $i\alpha$ of a non-zero root α is again a root for $i \neq 0, \pm 1$.

Proof. We can find in $L_{-\alpha}$ a non-zero element x characteristic under all of H. (As in the preceding proof, Lie's theorem could be cited, but since we know H to be abelian, elementary linear algebra as in Theorem 25 will do.) Since L_α and $L_{-\alpha}$ are dual relative to the Killing form (Theorem 46) we can find $y \in L_\alpha$ with $(x, y) \neq 0$. Form V, the subspace of L spanned by x, H, and $L_{i\alpha}$ for all positive integers i. Then V is invariant under R_x, R_y and $R_h = R_x R_y - R_y R_x$, where $h = xy$. The fact that the trace of R_h on V is 0 gives us the equation

$$(-1 + \Sigma\, id_i)\, \alpha(h) = 0 \ ,$$

where d_i is the dimension of $L_{i\alpha}$. If we know that $\alpha(h) \neq 0$, we can deduce $-1 + \Sigma\, id_i = 0$, which yields at once both conclusions of the theorem. So it remains to show that $\alpha(h) \neq 0$. We assume $\alpha(h) = 0$. Then by Theorem 51, $h = 0$. Pick $k \in H$ with $\alpha(k) \neq 0$. Then $(h, k) = (xy, k) = -(y, xk) = \alpha(k)(x, y) \neq 0$, a contradiction.

At this point we pause to summarize what we know of the structure of L, a semi-simple Lie algebra over an algebraically closed field of characteristic 0. We have a vector space direct sum decomposition

$$L = H + L_\alpha + L_{-\alpha} + L_\beta + L_{-\beta} + \ldots$$

with H abelian and the spaces L_α all one-dimensional. An element $h \in H$ acts on L_α as a multiplication by $\alpha(h)$, the latter being a linear function on H. The product $L_\alpha L_\beta$ falls into $L_{\alpha+\beta}$, this being interpreted as 0 if $\alpha+\beta$ is not a root; $L_\alpha L_{-\alpha} \subset H$. The Killing form is non-singular when restricted to H and pairs L_α, $L_{-\alpha}$ in a non-singular duality.

We next take a useful step made possible by the self-duality that the Killing form creates in H. For any root α there is a unique element h_α in H such that taking inner product with h_α induces the function α: $(h_\alpha, k) = \alpha(k)$ for any k in H. Note that

$$\beta(h_\alpha) = \alpha(h_\beta) = (h_\alpha, h_\beta).$$

In the theorems that follow we use the notation we have established without further reference.

Theorem 53. For $x \in L_\alpha$, $y \in L_{-\alpha}$ we have $xy = -(x, y)h_\alpha$.

Proof. For any k in H we have $(xy, k) = -(y, xk) = -\alpha(k)(x, y)$. This identifies xy with $-(x, y)h_\alpha$.

We note that it follows from Theorem 51 that $\alpha(h_\alpha) = (h_\alpha, h_\alpha)$ is not 0. We now prove more.

Theorem 54. For all α and β, (h_α, h_β) is rational. For all α, (h_α, h_α) is a positive rational number.

Proof. By Theorem 50, (h_α, h_β) is a rational multiple of (h_α, h_α). Thus the first statement follows from the second.

Write $(h_\alpha, h_\beta) = r(\alpha, \beta)(h_\alpha, h_\alpha)$, and let us write h for h_α to simplify notation. We revert to the definition of the Killing form to recall that $(h, h) = \mathrm{Tr}(R_h^2)$. Now R_h annihilates H, and on the one-dimensional space L_β it acts as a multiplication by $\beta(h)$. Hence

$$(h, h) = \Sigma\, \beta(h)^2 = \Sigma\, r(\alpha, \beta)^2 (h, h)^2\ ,$$

the sum being taken over all roots β. Since $(h,h) \neq 0$, this equation shows that (h,h) is a positive rational number.

Theorem 55. The h_α's span H.

Proof. If the h_α's span a proper subspace of H then there exists a non-zero element k in H orthogonal to every h_α. But this means $\alpha(k) = 0$ for every root, whence k lies in the center of L, a contradiction of semi-simplicity.

We write H_0 for the vector space over the rational numbers spanned by the h_α's.

Theorem 56. The Killing form restricted to H_0 is positive definite.

Proof. Let h be a non-zero element of H_0. By Theorem 54, $\alpha(h)$ is rational for every root α. This shows that R_h may be written as a diagonal matrix with rational entries on the diagonal. If $h \neq 0$ the matrix R_h is not 0. Hence $(h,h) = \mathrm{Tr}(R_h^2)$ is a positive rational number.

Theorem 57. The dimension of H_0 (over the rational numbers) is equal to the dimension of H.

Proof. Let $h_1, \ldots, h_r$ be a basis of H_0 over the rational numbers. The matrix of inner products (h_i, h_j) is positive definite and non-singular. This shows that the elements $h_1, \ldots, h_r$ continue to be linearly independent over the larger base field of L. By Theorem 55 they are a basis of H.

We proceed to get more information concerning the multiplicative relations between the L_α's. For this purpose the main tool is a theorem about representations of the simple three-dimensional Lie algebra.

Since Theorem 58 will be used again in §8, we formulate it so as to be valid for any characteristic.

Theorem 58. Let A, X and H be linear transformations on a vector space V. Assume that $AH - HA = A$, $XH - HX = -X$, $AX - XA = H$. Let u be a vector satisfying $uX = 0$, $uH = \lambda u$, λ a scalar. Then for all i,

(6) $$uA^iH = (\lambda + i)uA^i$$

(7) $$uA^iX = [i\lambda + (1/2)i(i-1)]uA^{i-1} .$$

Suppose that $uA^r = 0$ for a certain positive integer r, and let r be the smallest such integer. If the characteristic is $p \neq 0$, assume r prime to p. Then $\lambda = -(r-1)/2$ (where r is to be interpreted as an integer mod p in case the characteristic is p).

Remark. For $p = 2$, the hypothesis implies that r is odd, whence $r-1$ is even and $(r-1)/2$ is meaningful.

Proof. To prove (6) by induction, we note that
$$uA^iH = uA^{i-1} \cdot AH = uA^{i-1}(HA + A)$$
$$= (\lambda + i-1)uA^{i-1} \cdot A + uA^i = (\lambda + i)uA^i .$$
Then we can handle (7) by induction:
$$uA^iX = uA^{i-1}(XA + H)$$
$$= [(i-1)\lambda + (1/2)(i-1)(i-2)]uA^{i-2} \cdot A + (\lambda + i-1)uA^{i-1}$$
$$= [i\lambda + (1/2)i(i-1)]uA^{i-1} .$$
For the final statement of the theorem, we apply (7) with $i = r$ and observe that the left side vanishes while the element uA^{r-1} on the right side is non-zero. Hence the coefficient on the right side vanishes:

$$r\lambda + r(r-1)/2 = 0.$$

Our hypothesis permits us to cancel r, and we conclude that $\lambda = -(r-1)/2$.

We return to the setup of our Lie algebra L (semi-simple over an algebraically closed field of characteristic 0) and its Cartan decomposition. To apply Theorem 58 conveniently, we normalize h_α by replacing it by $h = h_\alpha/(h_\alpha, h_\alpha)$. The effect of this change is to make R_h act as the identity on L_α, and as the negative of the identity on $L_{-\alpha}$. Pick $a \in L_\alpha$, $x \in L_{-\alpha}$ to satisfy $ax = h$. Then the triple $A = R_a$, $X = R_x$, $H = R_h$ satisfies the commutation relations set out in Theorem 58. Now suppose that u is a non-zero element of L_β satisfying $ux = 0$. We have $uH = \lambda u$ where $\lambda = (h_\alpha, h_\beta)/(h_\alpha, h_\alpha)$. Theorem 59 is now immediate from Theorem 58.

Theorem 59. Let u, x, a be non-zero elements of $L_\beta, L_\alpha, L_{-\alpha}$. Suppose that $ux = 0$, and let r be the smallest integer such that $uR_a^r = 0$. Then

(8)
$$r - 1 = -2(h_\alpha, h_\beta)/(h_\alpha, h_\alpha).$$

It should be noted that in Theorem 59 we assumed $L_\beta L_{-\alpha} = 0$ but we did not assume that $\beta - \alpha$ is not a root. In other words, there might conceivably be a root space $L_{\beta-\alpha}$ waiting for the product of L_β and $L_{-\alpha}$, with the product perversely being zero. However we can now clear up this ambiguity.

Suppose we have the setup of Theorem 59 and that $\beta - \alpha$ is a root. Continue the sequence $\beta - \alpha, \beta - 2\alpha, \ldots$ back as far as possible, say to $\beta - m\alpha = \gamma$. Then any non-zero element y in L_γ satisfies $yx = 0$ and can play the role that u did above. If we write s for the smallest integer such that $yR_a^s = 0$, then Theorem 59 gives us

$$(9) \qquad s - 1 = -2(h_\alpha, h_\gamma)/(h_\alpha, h_\alpha).$$

Since $h_\gamma = h_\beta - mh_\alpha$ we deduce from (8) and (9) that $s = r + 2m$. But on the other hand yR_a^m is a multiple of u, whence $yR_a^{m+r} = 0$, and $s \leq m + r$, a contradiction. We have proved (after a change of notation):

Theorem 60. If α, β and $\alpha + \beta$ are all non-zero roots then $L_\alpha L_\beta = L_{\alpha + \beta}$.

We can now re-interpret Theorem 59 in a sharper way. Take roots β and α such that $\beta - \alpha$ is not a root. We run the roots $\beta, \beta + \alpha, \beta + 2\alpha, \cdots$ as far as possible, ending at $\beta + (r-1)\alpha$. Then r is determined by (8). We restate this as a formal theorem, at the same time replacing r by $r + 1$.

Theorem 61. Let α and β be roots with $\beta - \alpha$ not a root. Then $r = -2(h_\alpha, h_\beta)/(h_\alpha, h_\alpha)$ is a non-negative integer. It has the property that $\beta + \alpha, \beta + 2\alpha, \ldots, \beta + r\alpha$ are all roots and $\beta + (r+1)\alpha$ is not a root.

6. The geometric classification

In this section we shall abstract and discuss the class of geometric configurations we encountered in §5. It is desirable to have a brief name, and we call them C-systems ("C" for Cartan).

Definition. Let K be an ordered field and let V be a finite-dimensional vector space over K carrying a positive definite inner product which we denote by (,). By a C-system Γ in V we mean a finite set of non-zero vectors in V (we use $\alpha, \beta, \ldots$ for members of Γ) satisfying:

(1) Γ spans V (this is of course merely a harmless normalization),

(2) $\alpha \in \Gamma$ implies $-\alpha \in \Gamma$,

(3) If $\alpha \in \Gamma$ no integral multiple of α other than $\pm \alpha$ lies in Γ,

(4) If $\alpha, \beta \in \Gamma$ with $\beta \neq \pm \alpha$ and $\beta - \alpha \notin \Gamma$, then $r = -2(\alpha, \beta)/(\alpha, \alpha)$ is a non-negative integer,

(5) For this integer r we furthermore have that $\beta + \alpha, \beta + 2\alpha, \ldots,$ $\beta + r\alpha$ all lie in Γ but $\beta + (r+1)\alpha$ does not.

We maintain tradition by calling the elements of a C-system "roots", although this is now meaningless.

The developments in the last section can be summarized in the assertion that the h_α's derived from a Cartan subalgebra (of a semisimple Lie algebra over an algebraically closed field of characteristic 0) span a C-system over the field of rational numbers.

Let us begin the discussion by extending to any pair of roots the property that $2(\alpha, \beta)/(\alpha, \alpha)$ is integral. We move backwards in the sequence $\beta - \alpha, \beta - 2\alpha, \ldots$ till we reach, say, $\gamma = \beta - m\alpha$ with $\gamma - \alpha$ not a root. Then by axiom (4), $2(\alpha, \gamma)/(\alpha, \alpha)$ is integral, and from this we deduce that $2(\alpha, \beta)/(\alpha, \alpha)$ is integral. This is of course also true for $\beta = \pm \alpha$.

Multiply together the integers $2(\alpha, \beta)/(\alpha, \alpha)$ and $2(\alpha, \beta)/(\beta, \beta)$. We deduce that

$$4(\alpha, \beta)^2/(\alpha, \alpha)(\beta, \beta)$$

is integral. By the Schwartz inequality this is in absolute value at most 4. Hence we have the next two theorems.

Theorem 62. For any roots α and β in a C-system,
$2(\alpha, \beta)/(\alpha, \alpha) = 0, \pm 1, \pm 2,$ or ± 3.

Theorem 63. Suppose in a C-system that $\beta, \beta + \alpha, \ldots, \beta + r\alpha$ are all roots. Then $r \leq 3$.

We add an additional point.

Theorem 64. If α, β lie in a C-system and $\beta = c\alpha$ with c a scalar in the underlying field, then $c = \pm 1$.

Proof. From the fact that $2(\alpha, \beta)/(\alpha, \alpha)$ and $2(\alpha, \beta)/(\beta, \beta)$ are integers we get that $2c$ and $2/c$ are both integral. This implies that $|c| = 1, 2$ or $1/2$. But the possibilities 2 and $1/2$ are ruled out by axiom (3).

We bring into the discussion the reflection in the hyperplane perpendicular to α, noting that it is given by

$$\gamma \rightarrow \gamma - \frac{2(\alpha, \gamma)}{(\alpha, \alpha)} \alpha ,$$

γ ranging over V.

Consider a "maximal string" of roots $\beta, \beta + \alpha, \ldots, \beta + r\alpha$, so that $\beta - \alpha$ and $\beta + (r+1)\alpha$ are not roots. We have $r = -2(\alpha, \beta)/(\alpha, \alpha)$. It is a straightforward computation that $\beta + i\alpha$ is sent into $\beta + (r-i)\alpha$ by the reflection in the hyperplane perpendicular to α. Now any root can be inserted in some maximal string. Hence:

Theorem 65. Let Γ be a C-system, α a root. Then Γ is sent into itself by the reflection in the hyperplane perpendicular to α.

Before embarking on the next stage of the discussion, we survey what the problem is like in elementary geometric terms. Assume the underlying field to be the field of real numbers. We are thus given a finite number of non-zero vectors in a Euclidean space V. The fact that $4(\alpha, \beta)^2/(\alpha, \alpha)(\beta, \beta)$ is an integer (necessarily in the range -4 to 4) tells us that the acute angle between any two vectors is $30^{\circ}, 45^{\circ}, 60^{\circ}$ or 90°. In addition the configuration possesses a high degree of symmetry: it admits the reflection in the hyperplane perpendicular to any of the vectors.

Now when V is 2-dimensional, we instantly see what the possibilities are: the vectors must be spaced regularly around a circle with an intervening angle of $30^{\circ}, 45^{\circ}, 60^{\circ}$, or 90°. (The corresponding Lie algebras are G_2, B_2, A_2, and the direct sum of two copies of A_1.) In brief our problem is this: generalize this intuitive geometric argument, so obvious in the plane, to any dimension.

The classification of C-systems is accomplished by relating them to a simpler class of systems. This remarkably effective idea goes back to the earliest work on Lie algebras (that of Killing and Cartan).

Definition. Let K be an ordered field and let V be a finite-dimensional vector space over K carrying a positive definite inner product (,). By a D-system Δ in V we mean a basis of V such that for any $\alpha \neq \beta$ in Δ, $-2(\alpha, \beta)/(\alpha, \alpha)$ is a positive integer or 0.

Definition. Let Γ be a C-system, and let Δ be a D-system contained in Γ. (It is of course tacitly understood that Γ and Δ span the same vector space.) We say that Δ is nicely embedded in Γ if for every $\alpha \epsilon \Gamma$ the expression of α in terms of the members of Δ has integral coefficients either all ≥ 0 or all ≤ 0.

Remark. As compared with some presentations, we have preferred to state the definition of a D-system in terms of intrinsic properties and leave the lexicographic ordering as a tool in the existence proof.

Theorem 66. Any C-system possesses a subset which is a nicely embedded D-system.

Proof. Let Γ be the given C-system in the vector space V. Pick any basis of V and order V lexicographically relative to this basis (an element is positive if and only if the first non-vanishing coefficient is positive). As is traditional, we call a root (= element of Γ) simple if it is positive and cannot be expressed as a sum of two positive roots. (It is to be carefully noted that the decision as to whether a given root is simple depends upon the choice of the lexicographic ordering.) We claim that the set Δ of simple roots constitutes a nicely embedded D-system. The things that need to be verified will be treated under numbered headings.

(1) Let α and β be distinct simple roots. We claim that $\gamma = \alpha - \beta$ is not a root. For if it is positive, in the lexicographic ordering, then $\alpha = \beta + \gamma$ contradicts the simplicity of α. If γ is negative, $\beta = \alpha + (-\gamma)$ is a similar contradiction.

(2) It follows that if α and β are distinct simple roots, $-2(\alpha, \beta)/(\alpha, \alpha)$ is a non-negative integer.

(3) Let δ be a positive root. We show that δ is a positive integral combination of simple roots. If δ is itself simple, all is well. So suppose that $\delta = \delta_1 + \delta_2$ with δ_1, δ_2 positive. We argue by induction that the result is known for δ_1, δ_2 (we have a finite number of positive roots, they form a linearly ordered set, and we can base an induction on

the number of steps to the bottom). Hence we have the result for δ.

Given any root δ in Γ, either δ or $-\delta$ is positive. Hence either δ or $-\delta$ is a positive integral linear combination of simple roots.

(4) Since Γ spans V, it follows at once that Δ spans V.

(5) The one point remaining is that the simple roots are linearly independent. If there is a dependence relation between simple roots, we have an equation of the form

(10) $$\Sigma\, c_i \alpha_i = \Sigma\, d_j \beta_j$$

where the α's and β's are simple roots and the c's and d's are positive elements of K. Write γ for the common value of the two sides of (10). We have $\gamma \neq 0$ since γ is positive in the ordering of V. Hence (γ, γ) is positive. But this gives us the contradiction

$$0 < (\gamma, \gamma) = \Sigma\, c_i d_j (\alpha_i, \beta_j) \leq 0 \; .$$

This completes the proof of Theorem 66.

Now that we have proved the existence of nicely embedded D-systems, we shall not find it necessary to invoke again the lexicographic ordering. (However no real generalization has occurred: see Ex. 4).

Let Γ be a C-system, and Δ a D-system nicely embedded in Γ. In an obvious sense we can speak of the elements of Γ as being positive or negative (relative to Δ). In constructing Δ in Theorem 66 we picked out the elements of Δ as being indecomposable among the positive elements. We next show that this is true of any nicely embedded D-system, and we prove a slightly stronger statement.

Theorem 67. Let $\Delta = \{\alpha_1, \ldots, \alpha_n\}$ be a D-system nicely embedded in a C-system Γ. Let β be a positive element of Γ, so that

$\beta = \Sigma\, m_i \alpha_i$ with the m_i's integers ≥ 0. Then for at least one α_i we have $\beta - \alpha_i \in \Gamma$.

Proof. Assume the contrary. Then $(\beta, \alpha_i) \leq 0$ for all i and we deduce

$$(\beta, \beta) = (\beta, \Sigma\, m_i \alpha_i) \leq 0,$$

a contradiction.

In the next two theorems we show that the classification of C-systems reduces completely to that of D-systems.

Theorem 68. Let Γ be a C-system in the vector space V, Δ a nicely embedded D-system in Γ; let Γ_o, Δ_o be similarly given in V_o. Let f be an isometry of V onto V_o which carries Δ onto Δ_o. Then f carries Γ onto Γ_o.

Proof. Let $\Delta = \{\alpha_1, \ldots, \alpha_n\}$ and write $f(\alpha_i) = \beta$. Let $\gamma = \Sigma\, m_i \alpha_i$ be a positive element of Γ, so that the m_i's are integers ≥ 0. It evidently suffices for us to prove that $\gamma_o = \Sigma\, m_i \beta_i$ lies in Γ_o. We do this by induction on $\Sigma\, m_i$, and for brevity let us call this the level of γ. For level 1 the result is immediate. By Theorem 67, $\gamma - \alpha_i \in \Gamma$ for some α_i. By induction on the level, $\gamma_o - \beta_i \in \Gamma_o$. We move back in the chain $\gamma - \alpha_i, \gamma - 2\alpha_i, \ldots, \gamma - r\alpha_i$ as far as possible while staying in Γ. Again by induction on the level, the longest chain $\gamma_o - \beta_i, \gamma_o - 2\beta_i, \ldots, \gamma_o - r\beta_i$ in Γ_o has the same length. If we write $\varepsilon = \gamma - r\alpha_i$, then $\varepsilon - \alpha_i \notin \Gamma$ and we know that $-2(\alpha_i, \varepsilon)/(\alpha_i, \alpha_i)$ is a positive integer at least equal to r (since α_i may be added to ε at least r times while staying in Γ). Then with $\varepsilon_o = \gamma_o - r\beta_i$, $-2(\beta_i, \varepsilon_o)/(\beta_i, \beta_i)$ is the same positive integer. The fact that it is at least r proves that $\gamma_o \in \Gamma_o$.

Theorem 69. Let Δ and Δ_1 be D-systems both nicely embedded in the C-system Γ. Then there exists an isometry of Γ onto itself which carries Δ into Δ_1.

Proof. Let P and P_1 denote the positive elements in Γ relative to Δ and Δ_1 respectively. If $P = P_1$, of course $\Delta = \Delta_1$, so we assume $P \neq P_1$. Let $\alpha_1, \ldots, \alpha_n$ be the members of Δ. We shall show that application to Δ_1 of a reflection in one of the α_i's increases the size of $P \cap P_1$. There must be some α_i not in P_1. Let S be the reflection in the hyperplane perpendicular to α_i; note that S sends $-\alpha_i$ into α_i. Now $-\alpha_i \in P_1$, so that $\alpha_i \in P_1 S$. Take any $\beta = c_1 \alpha_1 + \ldots + c_n \alpha_n \in P_1 \cap P$. Then since

$$\beta S = \beta - \frac{2(\alpha_i, \beta)}{(\alpha_i, \alpha_i)} \alpha_i \ ,$$

we see that βS has at least one positive coefficient, so βS still lies in P, $\beta S \in P_1 S \cap P$. We have increased $P_1 \cap P$ by at least one in the passage to $P_1 S \cap P$, for we have kept all the old members of $P_1 \cap P$ and introduced the new member α_i.

We shall omit the proofs of the three remaining theorems. Among references that can be consulted are [23, Exposé 13], [11, pp. 128-135], [24, Ch. V] and [5].

Let Γ be a C-system in V. If there exists an orthogonal decomposition $V = V_1 \oplus V_2$ into subspaces in such a way that Γ is a set-theoretic union $\Gamma = \Gamma_1 \cup \Gamma_2$ with $\Gamma_i \subset V_i$ we say that Γ has been decomposed into Γ_1 and Γ_2. Necessarily Γ_i and Γ_2 are C-systems (in V_1 and V_2 respectively). If Γ has no decomposition we say that it is indecomposable. Entirely analogous definitions are made for D-systems.

Theorem 70. Any C-system has a unique decomposition into inde-composable C-systems. Any D-system has a unique decomposition into indecomposable D-systems.

Theorem 71. Let the D-system Δ be nicely embedded in the C-system Γ. Then Γ is indecomposable if and only if Δ is indecomposable.

At this point it is clear that the classification of C-systems has been reduced to that of indecomposable D-systems.

The visualization of indecomposable D-systems is greatly aided by a graphical representation due to Coxeter. Let $\Delta = \{\alpha, \beta, \ldots\}$ be a D-system. Construct a graph by inventing a vertex for every member of Δ. Join α and β by k lines where $k = 4(\alpha, \beta)^2/(\alpha, \alpha)(\beta, \beta)$. If K is the field of real numbers (so that we can speak of the angle between α and β), the number of lines is related to the angle as follows:

Angle	Number of lines
90°	0
120°	1
135°	2
150°	3

It is also necessary to pay a little attention to the lengths of the vectors (it turns out that this is needed only to distinguish B_n and C_n below; symmetry makes the lengths irrelevant in F_4 and G_2). Suppose that $4(\alpha, \beta)^2/(\alpha, \alpha)(\beta, \beta) = 2$. We readily see that one of (α, α) and (β, β) is twice the other. If (β, β) is the larger we put an arrow running from α to β on one of the two lines joining α and β:

$$\alpha \Longleftrightarrow \beta$$

63

We make the easy observation that a D-system is indecomposable if
and only if its Coxeter graph is connected, and we are ready to state
Theorem 72.

Theorem 72.   The following graphs give precisely all indecomposable D-systems up to multiplication by a constant:

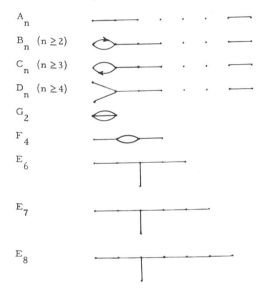

A_n

B_n $(n \geq 2)$

C_n $(n \geq 3)$

D_n $(n \geq 4)$

G_2

F_4

E_6

E_7

E_8

Remark. In every case the subscript gives the number of vertices.
The restrictions to $n \geq 2, 3, 4$ in B_n, C_n, D_n avoid duplications.

Questions of existence still remain. In fact, existence questions
can be posed on three levels:

(a) Existence of the D-systems. This is quite easy.

(b) Existence of the corresponding C-systems. This is somewhat
more difficult.

(c) Existence of the corresponding Lie algebras. This is still more difficult.

Full details are given in Jacobson [11].

Exercises

1. Let α be any root in a C-system. Show that α lies in a nicely embedded D-system.

2. Let Δ be a D-system. Show that the set of vectors $-\alpha$, where α runs over Δ, forms a D-system.

3. Let the D-system Δ be nicely embedded in the C-system Γ. Show that the set of negatives of members of Δ also forms a D-system nicely embedded in Γ.

4. Let the D-system Δ be nicely embedded in the C-system Γ in the vector space V. Prove that there exists a lexicographic order on V such that Δ is exactly the set of simple roots in Γ.

7. Transition to a geometric problem (characteristic p)

In this section and the next we shall develop the theory of suitable Lie algebras of characteristic p to the point where there emerge geometric systems analogous to the C-systems of §6.

The basic investigation of this type was carried out by Seligman [19]. In order to be able to state his fundamental axiom we have to define restricted representations.

Theorem 73. Let A be any algebra over a field of characteristic p, and let D be a derivation of A. Then D^p is a derivation of A.

Proof. By Leibnitz's rule we have, for any $x, y \in A$,

(11) $\quad (xy)D^p = xD^p \cdot y + pxD^{p-1} \cdot yD + \ldots + \binom{p}{i} xD^{p-i} \cdot yD^i + \ldots + x \cdot yD^p$.

Since the characteristic is p, all the intermediate binomial coefficients in (11) vanish. Hence we have

$$(xy)D^p = xD^p \cdot y + x \cdot yD^p ,$$

that is, D^p is a derivation.

We shall find it sufficient for our purposes to discuss the concept of restricted Lie algebras only for centerless ones (i.e., Lie algebras with center 0).

Definition. Let L be a centerless Lie algebra over a field of characteristic p. We say that L is restricted if the p-th power of any inner derivation is inner.

If L is restricted and $x \in L$, then there is a unique element in L which induces the same inner derivation as the p-th power of the inner derivation by x. We write x^p for this element. Its characteristic property is

(12) $\qquad yx^p = (\ldots(yx \cdot x)\ldots x)$

for all $y \in L$, there being p of the x's on the right side of (12).

We say that a representation S of a restricted Lie algebra L is restricted if $S(x^p) = (S(x))^p$ for all $x \in L$.

We can now describe the program of Seligman in [19]; he studied restricted Lie algebras admitting a restricted representation whose induced invariant form is non-singular.

We shall modify this program in two respects. First, following Block [1], we eliminate any hypothesis of restrictedness. In essence, it turns out that the relevant algebras are automatically restricted. While this is not true for the representations, by small additional arguments this point can be circumvented.

Second, we change the setup slightly by introducing "projective" representations. We present the motivation for this change. Let the characteristic be p. Let M be the algebra of all n by n matrices of trace 0, and suppose that p divides n. The center of M is the one-dimensional ideal Z of scalar matrices; note that $Z \subset M$ because p divides n. The algebra $N = M/Z$ is simple (with the solitary exception of the case $n = 2$, $p = 2$), and it is one of the algebras that deserves to appear at the end of a structure theorem. However it is proved in [2] that in any representation of N, the induced invariant form is identically 0.

Block and Zassenhaus [3], [4] coped with this difficulty by broadening the investigation so as to cover homomorphic images of Lie algebras admitting a suitable representation.

We prefer a more direct approach, which is in essence a special case of the work of Block and Zassenhaus. We begin with the observation that the above algebra $N = M/Z$ admits a splendid invariant form. To see this, take any X^* and Y^* in N and let $X, Y \in M$ map onto them. We propose to define $f(X^*, Y^*) = \text{Tr}(XY)$. To argue that this is well defined we change X by a member C of Z, and note that $\text{Tr}(CY) = 0$ since C is a scalar and Y has trace 0. That f is invariant and non-singular on N is a routine verification.

Remark. It is vital here that M be confined to matrices of trace 0. See the exercise at the end of this section.

By a projective representation of a Lie algebra L we mean a homomorphism of L into a Lie algebra of the type $N = M/Z$. There is an invariant form attached to any projective representation. Our program will be to study Lie algebras admitting a projective representation whose induced form is non-singular.

Before starting the theorems, we note that Seligman's algebras are included in ours. Let S be an ordinary representation of a Lie algebra L of characteristic p. Assume that $L^2 = L$. Enlarge each matrix $S(x)$ by bordering it with zeros till the size of the matrices is divisible by p. Divide by the scalars so as to pass to a projective representation. This projective representation gives rise to the same invariant form as the original representation S.

Theorem 74. Let $A \subseteq B$ be centerless Lie algebras such that B is restricted and B admits an invariant form which is non-singular on A. Then A is restricted.

Proof. We use the notation $(,)$ for the form. Fix x in A. The map $a \to (a, x^p)$ is a linear function on A which, since the form is non-singular on A, is induced by an element $c \in A$. We claim that c serves as the p-th power of x in A. To see this we have to show that multiplication by $c - x^p$ vanishes on A, and for this it suffices to prove that $((c - x^p)A, A) = 0$. This is true since the form is invariant and $(c - x^p, A) = 0$.

Theorem 75. Let L be a centerless Lie algebra of characteristic p. Assume that L admits a projective representation such that the induced invariant form is non-singular on L. Then L is restricted.

Proof. The representation is necessarily faithful. Thus we can look at L as a subalgebra of the algebra $N = M/Z$ into which we are given the projective representation (the notation $N = M/Z$ is being used as above). Now Theorem 74 is applicable.

Remark. It should be carefully noted that the representation in Theorem 75 need not be restricted.

Theorem 76. Let A and B be linear transformations on a finite-dimensional vector space V over a perfect field of characteristic p. Suppose that $AB - BA = I$ (the identity linear transformation), and that V is irreducible under A and B. Then V is p-dimensional and we have $A = \lambda I + A_1$, $B = \mu I + B_1$ where λ and μ are scalars and (relative to a suitable basis) A_1 and B_1 are given by

$$A_1 = \begin{pmatrix} 0 & 1 & 0 & \ldots & 0 \\ 0 & 0 & 1 & \ldots & 0 \\ & & \ldots & & \\ 0 & 0 & 0 & \ldots & 1 \\ 0 & 0 & 0 & \ldots & 0 \end{pmatrix}$$

$$B_1 = \begin{pmatrix} 0 & 0 & 0 & \ldots & 0 & 0 \\ 1 & 0 & 0 & \ldots & 0 & 0 \\ 0 & 2 & 0 & \ldots & 0 & 0 \\ & & & \ldots & & \\ 0 & 0 & 0 & \ldots & p-1 & 0 \end{pmatrix}$$

For $p > 2$ we have $\mathrm{Tr}(AB) = 0$.

Proof. We have $[[BA]A] = 0$, and so $[\ldots[BA]\ldots A] = 0$ where A is repeated p times. Hence $[BA^p] = 0$. Thus A^p commutes with A and B, and is a scalar by Schur's lemma. The p-th root of the scalar can be extracted. After subtracting off this scalar and changing notation we have $A^p = 0$. Similarly we normalize B so as to arrange $B^p = 0$.

Let x be a null vector for B. We compute $xAB = x$,

$$xA^2B = xA(BA + I) = 2xA, \text{ etc.}$$

till $xA^{p-1}B = (p-1)xA^{p-2}$. The vectors $x, xA, \ldots, xA^{p-1}$ are linearly independent, for if

$$c_o x + c_1 xA + \ldots + c_{p-1}xA^{p-1} = 0,$$

we apply B and get a similar shorter equation, ultimately reaching a contradiction. The subspace spanned by $x, xA, \ldots, xA^{p-1}$ is invariant under A and B, and hence is all of V. Relative to this basis we get the representation for A and B which is ascribed to A_1 and B_1 in the theorem.

We return to the notation A_1, B_1 and observe that

$$A_1 B_1 = \begin{pmatrix} 1 & 0 & 0 & \ldots & 0 & 0 \\ 0 & 2 & 0 & \ldots & 0 & 0 \\ & & \ldots & & & \\ 0 & 0 & 0 & \ldots & p-1 & 0 \\ 0 & 0 & 0 & \ldots & 0 & 0 \end{pmatrix}$$

so that

$$Tr(A_1 B_1) = 1 + 2 + \ldots + (p-1) = p(p-1)/2.$$

Hence for $p > 2$ we have $Tr(A_1 B_1) = 0$. Since

$$Tr(AB) = p\lambda\mu + \lambda Tr(B_1) + \mu Tr(A_1) + Tr(A_1 B_1)$$

and $Tr(A_1) = Tr(B_1) = 0$, we deduce that $Tr(AB) = 0$, as required.

We proceed to remove the assumption of irreducibility.

Theorem 77. Let A and B be linear transformations on a finite-dimensional vector space V over a field of characteristic $p > 2$. Assume that $AB - BA = I$, the identity linear transformation. Then $Tr(AB) = 0$.

Proof. The base field can be enlarged without changing the problem, and so we may assume it to be perfect (or indeed algebraically closed). It then suffices to decompose V into a composition series of irreducible spaces and quote Theorem 76.

The following is in essense a sharpening of Theorem 47.

Theorem 78. Let L be a nilpotent Lie algebra over an algebraically closed field of characteristic $p > 2$. Let a be an element of L^2 satisfying $aL \cdot L = 0$. Let S be an (ordinary) representation of L and f the resulting invariant form on L. Then $f(a, L) = 0$.

Proof. Since it is harmless to suppress the kernel of S, we can suppose that L is an algebra of linear transformations on a vector space V. We are given $A \in [LL]$ satisfying $[[AL]L] = 0$. For any $B \in L$ we have to prove that $Tr(AB) = 0$. It is clear that we may assume that V is irreducible under L.

Write $C = [AB]$. Then C is in the center of L. By Schur's lemma C is a scalar. There are two cases.

I. $C \neq 0$. It is harmless to assume that $C = I$. Then Theorem 77 tells us that $Tr(AB) = 0$.

II. $C = 0$, i.e., A and B commute. By Theorem 41, every element of L has the form scalar plus nilpotent. Since A and B commute, they can be put in simultaneous triangular form. If A is nilpotent, $Tr(AB) = 0$ is evident. If A is not nilpotent, we note that $Tr(A) = 0$ since $A \in [LL]$, and deduce that the dimension of V is divisible by p. Again $Tr(AB) = 0$ is evident.

Theorem 79. Let L_o be a nilpotent Lie algebra over an algebraically closed field of characteristic $p > 2$. Assume that L_o possesses a projective representation such that the induced invariant form is non-singular on L_o. Then L_o is abelian.

Proof. We may regard L_o as embedded in the algebra $N = M/Z$ in our usual notation. Let $L \subset M$ be the complete inverse image of L_o. Then there is a one-dimensional central C in L with $L/C = L_o$. Thus L is also nilpotent. On L we have an ordinary representation (namely the identity) with an induced invariant form for which we write f.

If L_o is not abelian there exists in L_o a non-zero element A^* which is in L_o^2 and in the center of L_o. Let $A \in L^2$ be an inverse image of A^*. Then we have $[AL] \subset C$, so that $[[AL]L] = 0$. By Theorem 78, $f(A, L) = 0$. Hence $(A^*, L_o) = 0$, where $(,)$ denotes the invariant form on N. This contradicts the assumed non-singularity of $(,)$.

Theorem 80. Let L be a Lie algebra over an algebraically closed field of characteristic $p > 2$. Assume that L possesses a projective representation such that the induced invariant form is non-singular on L. Let H be a Cartan subalgebra of L. Then H is abelian.

Proof. This follows from Theorems 46 and 79.

Theorem 81. Let L be a Lie algebra of linear transformations on a finite-dimensional vector space over an algebraically closed field of characteristic $p > 2$. Assume given $A \in L$, $B \in [LL]$ with the properties that $[AB]$ and $[A^p L]$ lie in the center of L. Then $\operatorname{Tr}(AB) = 0$.

Proof. We can assume V irreducible under L. Then by Schur's lemma, $C = [AB]$ is a scalar. If it is a non-zero scalar, Theorem 77 gives us the result. Hence we may assume that $C = 0$. Then A and B commute and may be put in simultaneous triangular form.

For any $D \in L$ we have that $[A^p D]$ is central. Hence commutating one more time with A gives 0. A fortiori, we have $[A^{p^2} D] = 0$. Thus A^{p^2} is central and is a scalar. It follows that A has the form scalar plus nilpotent. Since $B \in [LL]$ we have $Tr(B) = 0$ and again $Tr(AB) = 0$ is evident.

Theorem 82. Let L_0 be a Lie algebra over an algebraically closed field of characteristic $p > 2$. Assume that L_0 admits a projective representation such that the induced invariant form is non-singular on L_0. Assume further that L_0 is centerless and equal to its square. (Note that by Theorem 75, L_0 is restricted.) Let H be a Cartan subalgebra of L_0 and $a \in H$. Then $a^p = 0$ implies $a = 0$.

Proof. By Theorem 46, the form, which we write as $(,)$, is non-singular on H. It therefore suffices for us to prove $(a, b) = 0$ for any $b \in H$.

As in the proof of Theorem 79, we may regard L_0 as a subalgebra of $N = M/Z$, and we pass to the complete inverse image L of L_0. Let $A, B \in L$ be representatives of a and b. Since L_0 is equal to its square we can pick $B \in [LL]$. (We could also do this for A, but this is not needed.) We now argue that the hypotheses of Theorem 81 are fulfilled. For H is abelian by Theorem 80; hence $[AB]$ lies in the center of L. We picked B so that $B \in [LL]$. Finally, the hypothesis $a^p = 0$ translates to the fact that $[A^p L]$ is central.

Next we need an elementary result concerning the simultaneous diagonalization of matrices.

Theorem 83. Let S be a commutative linear space of matrices over an algebraically closed field of characteristic p. Assume that S is closed under the taking of p-th powers and that S contains no non-zero nilpotent matrices. Then S can be put into simultaneous diagonal form.

Proof. By standard linear algebra it suffices to prove that each individual matrix in S is diagonable. Let T denote the subset of diagonable matrices in S; T is evidently a subspace of S. The mapping $A \to A^p$ is a one-to-one additive mapping of T into itself (note that it is semi-linear rather than linear). We shall prove by a dimension argument that the mapping is onto. Let $B_1, \ldots, B_r$ be a basis of T. We claim that the matrices $B_1^p, \ldots, B_r^p$ likewise constitute a basis of T. It is enough to prove that they are linearly independent. Suppose that

$$c_1 B_1^p + \ldots + c_r B_r^p = 0.$$

Since the base field is perfect we may write $c_i = d_i^p$. We thus find $\Sigma \, d_i B_i = 0$, a contradiction.

Hence: every diagonable matrix in S is the p-th power of a diagonable matrix in S. Now for any $C \, \epsilon \, S$ we have that a suitable power C^{p^n} is diagonable. We write $C^{p^n} = D^{p^n}$ with D diagonable. But then $(C - D)^{p^n} = 0$, $C = D$. Theorem 83 stands proved.

Given a subalgebra S of a Lie algebra L, we shall for brevity say "S is diagonable" meaning that it is diagonable in the regular representation, that is, the set of linear transformations R_x, x ranging over S, admits simultaneous diagonal form.

We assemble in a single theorem the information given by Theorems 75, 80, 82, and 83.

Theorem 84. Let L be a Lie algebra over an algebraically closed field of characteristic p > 2. Assume that L is centerless and equal to its square, and that L admits a projective representation such that the invariant form induced on L is non-singular. Then L is restricted. Furthermore, any Cartan subalgebra of L is abelian and diagonable.

Exercise

Let L be the Lie algebra of all n by n matrices over a field of characteristic p. Assume that p divides n. Let Z be the center of L (i.e., the scalar matrices). Prove that any invariant form on L/Z vanishes on M/Z, where M consists of the matrices of trace 0.

8. Transition to a geometric problem (characteristic p), continued

In essence the material of the present section is a direct continuation of the preceding section. We have made a break in the exposition for three reasons.

(a) By starting fresh with a set of axioms we hope to make this section self-contained and clear.

(b) We have no further need to use the assumption that the form comes from a representation; it can be any invariant form.

(c) Treated this way, the Lie algebra can be allowed to be infinite-dimensional. This may be of interest since it allows the theory to encompass appropriate infinite-dimensional analogues of the classical

Lie algebras.

We now axiomatize the Lie algebras we shall treat. The exposition will carry their theory to the point where a well defined geometric problem arises.

We postulate the following elements of structure:

(1) An algebraically closed field F of characteristic $p > 3$, $p \neq 0$.

(2) A (possibly infinite-dimensional) Lie algebra L over F,

(3) A non-singular invariant form $(\ ,\ )$ on L,

(4) A vector space direct sum decomposition (possibly infinite) of L:

$$L = H \oplus L_\alpha \oplus L_\beta + \ldots$$

where H is abelian, $\alpha, \beta, \ldots$ are distinct non-zero linear functions on H, and we have $ah = \alpha(h)a$ for any $a \in L_\alpha$, $h \in H$.

We call the linear functions $\alpha, \beta, \ldots$ <u>roots</u> as usual. It helps to maintain uniform notation to write $H = L_0$. We call the dimension of H the <u>rank</u>, but this is not meant to prejudge any uniqueness questions. The spaces L_α are conceivably infinite-dimensional, but after we add a hypothesis we shall prove them to be one-dimensional. However, H and L will then still be possibly infinite-dimensional.

We call a Lie algebra equipped with all the above structure a <u>V-algebra</u> (V does not stand for anything in particular). We exclude the trivial case $L = H$.

To avoid excessive repetition it will be understood that in all the theorems of this section a V-algebra is under discussion.

Theorems 85 and 86 are really covered by Theorems 43 and 46, but we repeat them for completeness.

Theorem 85. $L_\alpha L_\beta \subset L_{\alpha+\beta}$ (where this is understood to include the statements $L_\alpha L_{-\alpha} \subset L_0 = H$, and $L_\alpha L_\beta = 0$ if $\beta \neq -\alpha$ and $\alpha+\beta$ is not a root).

Proof. For $a \in L_\alpha$, $b \in L_\beta$, $h \in H$ we have

$$ab \cdot h = a \cdot bh + ah \cdot b$$

by the Jacobi identity. Since $ah = \alpha(h)a$, $bh = \beta(h)b$, we deduce $ab \cdot h = (\alpha+\beta)ab$. This makes the theorem evident.

Theorem 86. If α is a root so is $-\alpha$. L decomposes into the following orthogonal direct sum:

$$L = H \oplus (L_\alpha \oplus L_{-\alpha}) \oplus (L_\beta \oplus L_{-\beta}) \oplus \ldots$$

The form is non-singular on H, vanishes on L_α, and makes L_α and $L_{-\alpha}$ into dual vector spaces.

Proof. We first prove $(H, L_\alpha) = 0$. Pick $k \in H$ with $\alpha(k) \neq 0$. Then R_k acts on L_α as a multiplication by $\alpha(k)$. Hence $L_\alpha k = L_\alpha$, and $(H, L_\alpha) = (H, L_\alpha k) = -(Hk, L_\alpha) = 0$.

Next we verify $(L_\alpha, L_\beta) = 0$ for $\beta \neq -\alpha$. With the same k as above, $(L_\alpha, L_\beta) = (L_\alpha k, L_\beta) = -(k, L_\alpha L_\beta)$, and this vanishes by Theorem 85 and the preceding paragraph.

The various statements of the theorem are now immediate.

If H were finite-dimensional, the linear function α on H would be induced by a unique vector of H (because the form is non-singular). Infinite-dimensionality of H forces us to be more cautious. But because of the extra information available in the present context there is such an element after all.

Theorem 87. For any root α there exists a (necessarily unique) element $h_\alpha \in H$ with $(h_\alpha, k) = \alpha(k)$ for all $k \in H$. For $a \in L_\alpha$, $x \in L_{-\alpha}$ we have $ax = -(a, x)h_\alpha$.

Proof. For $a \in L_\alpha$, $x \in L_{-\alpha}$, $k \in H$ we have

(13)
$$(ax, k) = -(x, ak) = -\alpha(k)(a, x).$$

Since L_α and $L_{-\alpha}$ are dually paired relative to the form (Theorem 86) we can pick a and x with $(a, x) = -1$. The resulting product ax furnishes the required element h_α, by (13). For general a and x we then deduce $ax = -(a, x)h_\alpha$, again by (13).

We proceed to the first major result: the demonstration that the only V-algebra of rank one is the simple three-dimensional Lie algebra.

Theorem 88. Let L be a V-algebra of rank one. Then L is simple and three-dimensional. There are two roots: a root α and its negative $-\alpha$. The root spaces L_α and $L_{-\alpha}$ are one-dimensional.

Proof. We remind the reader that the hypothesis that L has rank one means that H is one-dimensional.

Fix a root α. The corresponding element h_α spans H, so $\alpha(h_\alpha)$ must be non-zero. We can multiply the form (,) by a non-zero constant without changing anything essential, and we can therefore suppose $\alpha(h_\alpha)$ to be normalized as 1. We shorten the writing by replacing h_α by h, L_α by L_1. Likewise if there is a root space $L_{i\alpha}$ with i an integer (more exactly, an integer mod p), we similarly use the notation L_i. Observe that R_h acts on L_i as a multiplication by i. Observe further that for $a \in L_1$, $x \in L_{-1}$ we have $ax = -(a, x)h$ by Theorem 87.

The proof will now proceed by a sequence of lettered steps.

(a) If a and b are linearly independent in L_1, then $ab \neq 0$.

Since L_1 and L_{-1} are dually paired by the form we can find $x \in L_{-1}$ with $(a, x) = -1$, $(b, x) = 0$. Then $ax = h$, $bx = 0$. If $ab = 0$ we get a contradiction from the Jabcobi identity

$$ab \cdot x + bx \cdot a + xa \cdot b = 0.$$

(b) Let j be an integer in the range $(p+1)/2 \leq j \leq p-2$. Then there are no divisors of zero between L_{-1} and L_j. That is: if x and u are non-zero elements of L_{-1} and L_j, then $ux \neq 0$.

Suppose on the contrary that $ux = 0$. Pick $a \in L_1$ with $ax = h$. We apply Theorem 58 with R_a, R_x, and R_h playing the roles of A, X, and H. Note that $uH = ju$ so that λ is to be replaced by j. Iterated use of Theorem 85 shows that uA^{p-j} is a scalar multiple of h. We apply A two more times, getting a scalar multiple of a and then 0. Thus $uA^{p-j+2} = 0$. Let r be the smallest positive integer with $uA^r = 0$; we have $r \leq p-j+2$. We apply the last part of Theorem 58, obtaining $j \equiv -(r-1)/2 \pmod{p}$ or

(14) $$r \equiv 1 - 2j \pmod{p}.$$

In view of the restrictions on r and j we deduce from (14) the equation $r = 2p+1-2j$. Combining this with the inequality $r \leq p-j+2$ we get $p-1 \leq j$, contradicting the hypothesis on j.

(c) There are no divisors of zero between L_{-1} and L_2.

Reverse the roles, letting L_2 play the role of L_{-1}. Then L_{-1} will become $L_{(p+1)/2}$, and we see that (c) is a consequence of (b).

(d) If 2 is a root (i.e., $L_2 \neq 0$) then all integers from 1 to $p-1$ are roots.

It suffices to see that L_j is a root for j in the range $(p+1)/2 \leq j \leq p-3$. If this is not the case, take the smallest j such that j is a root and $j-1$ is not. We get a contradiction of (b).

(e) If any integer other than ± 1 is a root, then all integers from 1 to $p-1$ are roots.

If $(p+1)/2$ is a root we reverse the roles, treating $L_{(p+1)/2}$ as L_1. Then L_1 plays the role of L_2, and (d) applies. If $(p+1)/2$ is not a root, take j to be the smallest root with $(p+1)/2 < j \leq p-2$. Then $j-1$ is not a root, and we contradict (b).

(f) Let $x \in L_{-1}$. Then R_x induces a mapping from L_2 to L_1 which is one-to-one but not onto.

Part (c) says that is is one-to-one. Since $(L_2 x, x) = (L_2, xx) = 0$, we have that $L_2 x$ cannot be all of L_1 since it annihilates x, and the pairing of L_1 and L_{-1} is non-singular (Theorem 86).

(g) $\mathrm{Dim}(L_1) \leq 2$.

If $\mathrm{Dim}(L_1) \geq 3$, we can pick a and b linearly independent in L_1 and a non-zero element $y \in L_{-1}$ annihilating a and b. By the Jacobi identity, y annihilates ab. This contradicts parts (a) and (c).

(h) No integer other than ± 1 is a root.

If we suppose the contrary then, by part (e), all integers from 1 to $p-1$ are roots. By part (g), which is applicable to any root space, the spaces L_i all have dimension at most 2. By part (f) we have the strict inequalities

$$\mathrm{Dim}(L_1) > \mathrm{Dim}(L_2) > \mathrm{Dim}(L_4),$$

and this is a contradiction.

(i) $\mathrm{Dim}(L_1) = 1$.

This follows from part (a), and the fact (part (h)) that 2 is not a root.

(j) The only roots are ± 1.

It remains to see that we cannot have a root λ, where λ is a field element not an integer mod p. We suppose the contrary and pick $u \neq 0$ in L_λ. Select $a \epsilon L_1$, $x \epsilon L_{-1}$ with $ax = h$, and write A, X, H for R_a, R_x, R_h. We note that $uH = \lambda u$. We claim that $\lambda - 1$ is a root. If we deny this, then $uX = 0$ and Theorem 58 is applicable. Since λ is not an integer mod p, we conclude that $uA^i \neq 0$ for all i. In particular, $uA^{p-1} \neq 0$, showing that $\lambda - 1$ is a root after all. In the same way $\lambda + 1$ is a root. Now the ratio of $\lambda - 1$ to $\lambda + 1$ is not an integer mod p. Hence what we have just proved is applicable to see that $(\lambda + 1) - (\lambda - 1) = 2$ is a root, contradicting part (h).

With parts (i) and (j), the proof of Theorem 88 is complete.

We call a root α <u>isotropic</u> if $(h_\alpha, h_\alpha) = 0$, otherwise <u>non-isotropic</u>. Theorem 89, an easy consequence of Theorem 88, shows that non-isotropic roots are well behaved.

<u>Theorem 89.</u> Let α be a non-isotropic root in a V-algebra L. Then L_α is one-dimensional. No scalar multiple of α is a root, other than $\pm \alpha$.

<u>Proof.</u> We form the following subalgebra of L:

$$L_o = H_o + \Sigma L_\beta$$

where H_o is the one-dimensional subspace of H spanned by h_α, and the sum is taken over all roots β which are scalar multiples of α. It is immediate that L_o is a V-algebra of rank 1, and Theorem 89 is a consequence of Theorem 88.

We are anxious to have all roots non-isotropic. Since Ex. 7 shows that this is not true in all V-algebras, we have to add a hypothesis. Our choice for an additional axiom is that the V-algebra be centerless.

Theorem 90. In a centerless V-algebra all roots are non-isotropic.

Proof. The proof follows that of [12, Th. 6]; see also [21, Lemma II. 2. 9]. We divide it into a number of steps. Let α be isotropic.

We say that a root β is orthogonal to α if $(h_\alpha, h_\beta) = 0$.

(a) If a root β is not orthogonal to α then $\beta - \alpha$ is a root.

Pick $a \in L_\alpha$, $x \in L_{-\alpha}$ with $(a, x) = -1$ so that (Theorem 87) $ax = h_\alpha$. Since α is isotropic, ah_α and xh_α are both 0. Take $u \neq 0$ in L_β. Assume that $\beta - \alpha$ is not a root; then $ux = 0$. We now prove

(15) $$uR_a^i R_x = i\beta(h_\alpha)uR_a^{i-1}$$

for all i. For $i = 0$ both sides of (15) vanish. Assume (15) known for $i - 1$. Then

(16) $$uR_a^i R_x = uR_a^{i-1}(R_x R_a + R_{h_\alpha}) = (i-1)\beta(h_\alpha)uR_a^{i-1} + uR_{h_\alpha}R_a^{i-1} .$$

since R_a and R_h commute. Now $uR_{h_\alpha} = \beta(h_\alpha)u$, and so (16) yields (15). The element uR_a^{p-1} vanishes since $\beta - \alpha$ is not a root. Working down from $i = p-1$ in (15), and using the assumption that $\beta(h_\alpha) \neq 0$, we reach the contradiction $u = 0$.

(b) If a root β is not orthogonal to α, then β is non-isotropic.

Suppose on the contrary that β is isotropic. By part (a), $\alpha + \beta$ is a root. Since $(h_\alpha, h_{\alpha+\beta}) = (h_\alpha, h_\alpha + h_\beta) = (h_\alpha, h_\beta) \neq 0$, $\alpha + \beta$ is not orthogonal to α. Hence, again by part (a), $2\alpha + \beta$ is a root. Similarly, we argue that $2\alpha + \beta$ is not orthogonal to β, and hence $2\alpha + 2\beta$ is a root. But $(h_{\alpha+\beta}, h_{\alpha+\beta}) = 2(h_\alpha, h_\beta) \neq 0$. Thus $\alpha + \beta$ is a non-isotropic root. This contradicts Theorem 89.

(c) If a root β is not orthogonal to α, then $\alpha + 2\beta$ and $\alpha - 2\beta$ are not roots.

Since we can replace β by $-\beta$ it suffices to prove that $\alpha - 2\beta$ is not a root. If on the contrary $\alpha - 2\beta$ is a root, then it is not orthogonal to α, and by part (a), $2\alpha - 2\beta$ is a root. By part (a) again, $\alpha - \beta$ is a root. By part (b), $\alpha - \beta$ is non-isotropic, since it is not orthogonal to α. This contradicts Theorem 89.

(d) If β is a root not orthogonal to α, then $(h_\alpha, h_\beta)/(h_\beta, h_\beta)$ is not an integer mod p.

Note that β is non-isotropic by part (b), so that we may legally put (h_β, h_β) in the denominator. Suppose on the contrary that

$$(h_\alpha, h_\beta)/(h_\beta, h_\beta) \equiv i \pmod{p}.$$

Necessarily $i \not\equiv 0 \pmod{p}$. Take j with $1 \le j \le p-1$, $j \equiv -1/2i \pmod{p}$. By iterated use of part (a), $\gamma = \beta + j\alpha$ is a root. However

$$(h_\gamma, h_\gamma) = (h_\beta + jh_\alpha, h_\beta' + jh_\alpha)$$

is 0 because of the way we chose j. But by part (b), γ is non-isotropic, since it is not orthogonal to α.

We are ready to complete the proof of Theorem 90. There must exist some root δ not orthogonal to the given isotropic root α, for otherwise h_α would be central, contrary to our hypothesis that the algebra is centerless. By part (b), δ is non-isotropic. By parts (a) and (c), $\alpha - \delta$ is a root, but $\alpha - 2\delta$ and $\alpha + 2\delta$ are not roots. Write $h = h_\delta/(h_\delta, h_\delta)$. Pick $a \in L_\delta$, $x \in L_{-\delta}$ with $ax = h$. Note that $ah = a$, $xh = -x$. Pick $u \ne 0$ in $L_{\alpha-\delta}$. Then $ux = 0$, $uR_a^3 = 0$. With $A = R_a$, $X = R_x$, and $H = R_h$ we are ready to apply Theorem 58. The λ of Theorem 58 is $(h_\delta, h_{\alpha-\delta})/(h_\delta, h_\delta)$, which, by part (d), is not an integer mod p. But Theorem 58 tells us that $\lambda = -(r-1)/2$, where r is an integer which is in fact at most 3. This contradiction completes the proof of Theorem 90.

To squeeze a little more generality out of the remaining theorems of this section we add to the axioms for a V-algebra the assumption that all roots are non-isotropic. It should be noted that this makes Theorem 89 applicable to tell us that all root spaces are one-dimensional and that if α is a root then no scalar multiple of α other than $\pm\alpha$ is a root. In particular, if α is a root then 2α is not a root.

For a V-algebra L we have the implications

$$L^2 = L \implies L \text{ is centerless} \implies \text{roots non-isotropic.}$$

The second implication is of course Theorem 90. For the first see Ex. 2(c). Ex. 6 furnishes an example to show that the second implication cannot be reversed. The first implication also is irreversible (Ex. 5), although (**Ex.** 4) it takes an infinite-dimensional example to show this.

To avoid a lot of repetition we make the following blanket assumption: in Theorem 91-96 the roots in question lie in a V-algebra for which we assume that all roots are non-isotropic.

Theorem 91. Let α and β be roots with $\beta \neq \pm\alpha$. Suppose that $\beta + \alpha$ and $\beta - \alpha$ are not roots. Then $(h_\alpha, h_\beta) = 0$.

Proof. Pick $a \in L_\alpha$, $x \in L_{-\alpha}$ with $ax = h_\alpha$. Then $L_\beta a = 0$ and $L_\beta x = 0$. Hence $L_\beta L_\alpha = 0$. But h_α acts on L_β as a multiplication by $\beta(h_\alpha) = (h_\alpha, h_\beta)$. Hence $(h_\alpha, h_\beta) = 0$.

Theorem 92. Let α and β be any roots. It is not possible for all of $\beta - 2\alpha$, $\beta - \alpha$, β, $\beta + \alpha$, and $\beta + 2\alpha$ to be roots.

Proof. Assume the contrary. Observe that neither $(\beta - 2\alpha) + \beta$ nor $(\beta - 2\alpha) - \beta$ is a root. By Theorem 91, $(h_{\beta - 2\alpha}, h_\beta) = 0$. Likewise $(h_{\beta + 2\alpha}, h_\beta) = 0$. Adding, we get the contradiction $(h_\beta, h_\beta) = 0$.

Remark. Theorem 92 is the characteristic p analogue of Theorem 63. For lack of positive definiteness we had to give an alternate argument.

At this point the arguments that led to Theorems 60 and 61 can be repeated nearly verbatim. So we merely state the final four theorems of this section.

Theorem 93. If α, β, and $\alpha + \beta$ are roots, then $L_\alpha L_\beta = L_{\alpha + \beta}$.

Theorem 94. Let α, β be roots, $\beta \neq \pm\alpha$. Assume that $\beta - \alpha$ is not a root. Then $r = -2(h_\alpha, h_\beta)/(h_\alpha, h_\alpha) = 0, 1, 2$ or 3 (more exactly this is a congruence mod p). For this integer r we have that $\beta + \alpha, \ldots, \beta + r\alpha$ are roots and $\beta + (r+1)\alpha$ is not a root.

Theorem 95. For any roots α, β,

$$2(h_\alpha, h_\beta)/(h_\alpha, h_\alpha) = 0, \pm 1, \pm 2 \text{ or } \pm 3 .$$

Theorem 96. Let α, β be roots with $\beta \neq \pm \alpha$. Assume that $\beta, \beta + \alpha, \ldots, \beta + r\alpha$ are roots and that $\beta - \alpha$ and $\beta + (r+1)\alpha$ are not roots. Then $r = -2(h_\alpha, h_\beta)/(h_\alpha, h_\alpha)$.

The classification of these geometrical structures for the finite-dimensional case (characteristic $\neq 0$) was carried out by Seligman [19], and simplified by Seligman and Mills [22]. Mills [15] extended the work to characteristics 5 and 7. The results can be generalized to the case of countable dimension by the methods used by Schue in [18]. However, a different idea is definitely needed in the uncountable case. Such an idea has been supplied by Robert Kibler (unpublished as yet). His method also gives an alternate discussion for the finite-dimensional case which may be of interest, since, even as simplified by Seligman and Mills, the characteristic p case remains fairly formidable.

Exercises

1. If $L = H + \Sigma L_\alpha$ is a finite-dimensional V-algebra, prove that H is a Cartan subalgebra of L.

2. In a V-algebra L let H^* be the subspace spanned by the h_α's.

(a) Prove that $L^2 = H^* + \Sigma L_\alpha$.

(b) Prove that the center Z of L is the orthogonal complement of H^* within H.

(c) If $L^2 = L$, prove that $Z = 0$.

(d) If $Z = 0$ prove that L^2 is a V-algebra.

(e) Prove that $L^2 \cdot L^2 = L^2$.

3. Let L be a V-algebra with center Z. Prove that L/Z is centerless.

4. Let L be a centerless V-algebra with H finite-dimensional. Prove that $L = L^2$. (Hint: by Ex. 2(b), the orthogonal complement of H^* in H is 0. Since H is finite-dimensional, $H^* = H$.)

5. Let L be the Lie algebra of all infinite matrices with zeros in all but a finite number of entries. Give a decomposition that makes L a V-algebra. Prove that L is centerless but that $L \neq L^2$.

6. Let L be the Lie algebra of all n by n matrices over a field of characteristic p, where p divides n. Give a decomposition making L a V-algebra with non-isotropic roots. Note that $L \neq L^2$ and that L

has a one-dimensional center. Prove that neither L^2 nor L/Z is a V-algebra but that L^2/Z is.

7. Let L be the following 4-dimensional Lie algebra: basis a, x, h, k with $ax = h$, $ak = a$, $xk = -x$, all other products 0. Define a form by $(a, x) = -(h, k) = 1$, all other inner products 0. Show that L is a V-algebra. Observe that there is one root and that it is isotropic.

1. NSS groups

Definition.   A topological group has no small subgroups if there exists a neighborhood U of the identity element 1 such that the only subgroup in U is $\{1\}$. We abbreviate the name to NSS.

A major part of our program is to establish the equivalences

  ·locally compact NSS $\Leftrightarrow$ locally Euclidean $\Leftrightarrow$ Lie .

We recall that a locally Euclidean group is a topological group with a neighborhood of the identity homeomorphic to Euclidean space; in a Lie group the homeomorphism can moreover be picked so as to make the group operations analytic. Thus the implication Lie $\Rightarrow$ locally Euclidean is tautologous. That Lie $\Rightarrow$ NSS becomes apparent early in the theory of Lie groups. The remaining implications are decidedly non-trivial.

Examples of NSS groups.

1. Any discrete group (take $U = \{1\}$).

2. The real line (additive group of real numbers).

3. The circle group (reals mod 1 or complex numbers of absolute value 1).

4. The full linear group over the reals or complexes. (Sketch: take a neighborhood of the identity matrix I so small that matrices in

the neighborhood have characteristic roots very close to 1. Then the characteristic roots must be 1, for otherwise their powers will wander too far from 1. Such a matrix is of the form $I+N$, N nilpotent. For large k, $(I+N)^k = I+kN+\dots$ will not be close to I unless $N=0$.)

5. As a generalization of example 4, let F be a topological field in which the additive and multiplicative groups are NSS. (Remark: _e continuity of the inverse is not always required in a topological field, but it should be assumed here.) Then the full linear group over F is NSS.

6. As another generalization of example 4, the multiplicative group of invertible elements in a Banach algebra is NSS. (This is an immediate consequence of a theorem of Gelfand: Satz 1 in [7].)

We offer five easy exercises for the reader to try.

1. If an NSS group is torsion of bounded order, prove that it is discrete. Give an example to show that "bounded order" cannot be dropped.

2. Let G be a topological group and H a closed normal subgroup. If H and G/H are NSS, then G is NSS. (In due course, this will give us the theorem that if H and G/H are Lie, so is G. Before Hilbert's fifth problem was solved, it was a significant achievement to prove this.)

3. If $G_1, \dots, G_n$ are NSS, so is $G_1 \times \dots \times G_n$.

4. An infinite direct product of groups, each with at least two elements, is never NSS.

5. If a topological group G admits a continuous isomorphism into an NSS group, then G is NSS. (In due course, this will give us Cartan's theorem that if a locally compact group G admits a continuous isomorphism into a Lie group, then G is a Lie group.)

The following appears to be open: if G is NSS and H is a closed normal subgroup of G, is G/H NSS? This is true if in addition G is locally compact, but we shall only be able to prove it late in the game. (Of course it is an old result for Lie groups.)

2. Existence of one-parameter subgroups

We now embark on our lengthy study of the structure of locally compact groups, with special emphasis throughout on the NSS case.

We shall make heavy use of sequential arguments and repeated choices of subsequences, although this was frowned on by Gleason in [8]. Thus it is vital for us to have available the first axiom of countability. At several places we shall moreover wish to take advantage of metrizability. We therefore quote, without including a proof, the standard theorem that a topological group satisfying the first axiom of countability is metrizable.

Theorem 2 below asserts that any locally compact NSS group is metrizable. In Theorem 1 we prove a more general result which will be useful later in coping with non-metrizable groups.

First we present a definition.

Definition. A subset S of a group G is __symmetric__ if $x \in S$ if and only if $x^{-1} \in S$. If we denote by S^{-1} the set of all x with $x^{-1} \in S$, the condition for symmetry can be written $S = S^{-1}$.

Observe that in a topological group any neighborhood V of the identity contains a symmetric neighborhood of the identity, for example $V \cap V^{-1}$.

Theorem 1. Let G be a locally compact group which can be generated by a compact symmetric neighborhood U of 1. Then any neighborhood of 1 in G contains a compact normal subgroup N such that G/N is metrizable.

The proof of Theorem 1 will be preceded by two lemmas.

Lemma 1. In a topological group G let U be a neighborhood of 1 and let K be a compact subset of G. Then there exists a neighborhood V of 1 satisfying $x^{-1}Vx \subset U$ for all $x \in K$.

Proof. Fix x in K for the moment. Since $x^{-1}1x = 1$, by continuity there exist open neighborhoods W_x of x and V_x of 1 such that $y^{-1}ay \in U$ for all $y \in W_x$, $a \in V_x$. Now let x vary over K. A finite number of W_x's, say $W_1, \ldots, W_n$ cover K. If $V_1, \ldots, V_n$ are the corresponding V_x's, then $V = \bigcap V_i$ fulfils the requirements of the lemma.

Lemma 2. In a Hausdorff space, let $\{U_i\}$ be a decreasing sequence of compact neighborhoods of a point z. Suppose that $\bigcap U_i = \{z\}$. Then the U's form a neighborhood base at z.

Proof. Given an open neighborhood V of z, we must prove that some U_i is contained in V. Assume the contrary. Let F be the complement of V, so that F is closed. Then each $F \cap U_i$ is non-void. Hence $\bigcap (F \cap U_i)$ is non-void, a contradiction since $\bigcap U_i = \{z\}$ and $z \notin F$.

Proof of Theorem 1. Let V be the given neighborhood of 1. We must find $N \subset V$ with the requisite properties. We can suppose that V is compact and symmetric. We set $W_1 = V$ and proceed to construct

a certain sequence $\{W_i\}$ of compact symmetric neighborhoods of 1.
Suppose W_i has already been constructed. By Lemma 1 (with W_i and
U playing the roles of U and K respectively), we can find a neighborhood Y of 1 such that $x^{-1}Yx \subset W_i$ for all $x \in U$. By shrinking Y to
W_{i+1} we can further arrange that W_{i+1} is a compact symmetric neighborhood of 1 satisfying $W_{i+1}W_{i+1} \subset W_i$. Let $N = \bigcap W_i$. Then $N \subset V$,
and $x^{-1}Nx = N$ for all $x \in U$. Since U generates G, N is normal
in G. In the quotient group G/N, the images of the W's form a descending sequence of compact neighborhoods with intersection the identity. By Lemma 2, G/N is metrizable.

Since any locally compact group G possesses a compact symmetric neighborhood U of 1, and the subgroup generated by U is open
in G, Theorem 2 is an immediate corollary of Theorem 1.

Theorem 2.   Any locally compact NSS group is metrizable.

We shall henceforth freely use sequential arguments in our discussion of locally compact NSS groups.

Theorem 3.   Let G be a locally compact NSS group. Then there
exists in G a neighborhood U of 1 on which squaring is one-to-one
(i.e., $x, y \in U$ and $x^2 = y^2$ imply $x = y$).

Proof.   Assume the contrary. Then there exists sequences $x_i \to 1$,
$y_i \to 1$ with $x_i \neq y_i$ and $x_i^2 = y_i^2$. Write $a_i = x_i y_i^{-1}$. Then $a_i \neq 1$ and
$a_i \to 1$. We have $y_i^{-1} a_i y_i = a_i^{-1}$ and so $y_i^{-1} a_i^k y_i = a_i^{-k}$ for any k.

Fix a compact symmetric neighborhood W of 1 which contains no
subgroup $\neq 1$. It is not possible that all powers of a_i lie in W, for then
the subgroup generated by a_i would be contained in W. Let m_i be the

unique positive integer such that $a_i, a_i^2, \ldots, a_i^{m_i}$ all lie in W, but $a_i^{m_i+1}$ does not. Now the sequence $\{a_i^{m_i}\}$ lies in the compact set W. Hence there is a convergent subsequence. After a change of notation we can suppose that the entire sequence $\{a_i^{m_i}\}$ converges, say to b. We have $b \in W$. Multiplying by the sequence $\{a_i\}$, which converges to 1, we see that $a_i^{m_i+1}$ also converges to b. Since $a_i^{m_i+1} \notin W$, we have $b \neq 1$ (indeed b is on the boundary of W). Now proceed to the limit in the equation $y_i^{-1} a_i^{m_i} y_i = a_i^{-m_i}$. The result is $b = b^{-1}$, or $b^2 = 1$. Thus the two-element subgroup $\{1, b\}$ lies in W, a contradiction, since W was assumed to contain no subgroup $\neq 1$.

We shall wish to make consistent use of a neighborhood which contains no subgroups and in addition has the property asserted in Theorem 3. We make a definition.

Definition. Let G be a locally compact NSS group. By a _canonical neighborhood_ in G we mean a compact symmetric neighborhood of 1 which contains no subgroup $\neq 1$ and has the property that squaring is one-to-one on it.

In dealing with a locally compact NSS group, the letter U will henceforth be reserved for a canonical neighborhood; there will moreover be long arguments in which U will be kept fixed. (Remark: in Theorems 4 and 8, however, the hypothesis that squaring is one-to-one on U will not be used.)

We proceed to the construction of a one-parameter subgroup in a locally compact NSS group (a one-parameter subgroup in a topological group G is a continuous homomorphism of the additive group of real numbers into G).

A remark on notation: we shall use $[x]$ only when x is a real number ≥ 0; it denotes the largest non-negative integer $\leq x$.

Theorem 4 serves as a prelude to Theorem 5 and will be used again (in the version given in Theorem 8).

Theorem 4. Let U be a canonical neighborhood in a locally compact NSS group. Suppose given a sequence $\{a_i\}$ of elements of G, and positive integers $\{m_i\}$ such that, for all i,

$$a_i, a_i^2, \ldots, a_i^{m_i} \in U .$$

Let a neighborhood V of 1 be given. Then there exists a positive real number r_o such that $a_i^{[rm_i]} \in V$ for all i and all r with $0 \leq r < r_o$.

Proof. Suppose not. Then, after dropping to a subsequence and changing notation, we have a sequence $\{r_i\}$ of positive numbers, $r_i \to 0$, such that $a_i^{[r_i m_i]} \notin V$ for all i. From some point on $r_i \leq 1$ and so $a_i^{[r_i m_i]} \in U$. After passing to another subsequence, and another change of notation, we may assume that $a_i^{[r_i m_i]} \to b$. Since $a_i^{[r_i m_i]} \notin V$, we have $b \neq 1$. For any positive integer p, $b^p = \lim a_i^{p[r_i m_i]}$. Now

$$p[r_i m_i] \leq pr_i m_i \leq m_i$$

for large i. Hence $b^p \in U$. Since U is symmetric, $b^{-p} \in U$ as well. Thus the whole cyclic subgroup generated by b lies in U, a contradiction.

Theorem 5. Let U be a canonical neighborhood in a locally compact NSS group G. Suppose given a sequence $\{a_i\}$ of elements of G and positive integers m_i such that

$$a_i, a_i^2, \ldots, a_i^{m_i} \in U,$$

$a_i \to 1$, and $a_i^{m_i}$ converges, say to $X(1)$. Then for every $r \geq 0$, $a_i^{[rm_i]}$ converges, say to $X(r)$. The mapping $r \to X(r)$ is continuous and extends to a unique one-parameter subgroup in G.

Proof. We begin with an observation. Suppose that $a_i^{[rm_i]} \to b$ and $a_i^{[sm_i]} \to c$. Then, since $[(r+s)m_i]$ and $[rm_i] + [sm_i]$ differ by at most one, and $a_i \to 1$, we deduce that $a_i^{[(r+s)m_i]} \to bc$.

We prove that $a_i^{[rm_i]}$ converges for $r = 1/2$. Since all these elements lie in the compact set U, it suffices to prove that any two convergent subsequences have the same limit. Suppose that y and z are limits of two such convergent subsequences. Using the remark in the preceding paragraph, we see that $y^2 = X(1)$, $z^2 = X(1)$. By the uniqueness of squaring, $y = z$.

By iterating this argument we find that $a_i^{[rm_i]}$ converges for any r of the form $r = 1/2^k$. Using again the observation in the first paragraph we get convergence for any dyadic rational (rational number with denominator a power of 2).

We tackle a general r with $r \leq 1$. Once again we assume that subsequences of $\{a_i^{[rm_i]}\}$ converge to y and z, and have to prove $y = z$. Let W be any neighborhood of 1; we shall prove that $y^{-1}z \in W$. Pick a compact symmetric neighborhood V of 1 with $VV \subset W$, and let r_o be chosen to correspond to V as in Theorem 4. Let s be a dyadic rational satisfying $s < r$, $r-s < r_o$. Say $a_i^{[sm_i]} \to u$. To avoid cumbersome notation, suppose that $a_i^{[rm_i]}$ is already the sequence converging to y, and (another subsequence and change of notation !)

$a_i^{[(r-s)m_i]} \to v$. We have $v \in V$, and $y = uv$. In exactly the same way, $z = uv_o$, $v_o \in V$. Hence $y^{-1}z = v^{-1}v_o \in VV \subset W$, as required.

The rest of the proof of Theorem 5 is quite routine: we check convergence of $a_i^{[rm_i]}$ for every $r \geq 0$, say to $X(r)$, we define $X(-r)$ as $X(r)^{-1}$, and we verify the equation $X(r+s) = X(r)X(s)$. Continuity of $r \to X(r)$ at $r = 0$ follows from Theorem 4, and (as always for topological groups) continuity of X everywhere is a consequence. Theorem 5 stands proved.

Now suppose in addition that G is non-discrete. Then a sequence a_i with $a_i \neq 1$, $a_i \to 1$ can be chosen. With U canonical, let m_i be the largest positive integer with $a_i, \ldots, a_i^{m_i}$ all in U. If a subsequence of $\{a_i^{m_i}\}$ converges to b, we have that $b \neq 1$ and that a one-parameter subgroup X exists with $X(1) = b$. Hence:

Theorem 6. A non-discrete locally compact NSS group possesses a non-trivial one-parameter subgroup.

Theorem 7. Let U be a canonical neighborhood in a locally compact NSS group. Let K be the set of points of the form $X(1)$, where X is a one-parameter subgroup with $X(t) \in U$ for all $|t| \leq 1$. Then K is closed.

Proof. Let $c_i \in K$ be a convergent sequence converging to c. We have to prove that $c \in K$. We have one-parameter subgroups X_i with $X_i(1) = c_i, X_i(t) \in U$ for $|t| \leq 1$. Define $a_i = X_i(1/i)$ and $m_i = i$ so that $a_i^{m_i} = c_i$. Apply Theorem 5.

We shall eventually prove that K fills a neighborhood of 1 in G.

In concluding this section we shall introduce a definition, and re-cast Theorems 4 and 5 in its language. We could have proved Theorems 4 and 5 at once in this slight extra generality, but it perhaps makes the exposition a little pleasanter to break this discussion into two parts in this way. The use of "standard" sequences will enable later arguments to proceed a little more efficiently.

<u>Definition.</u> Let U be a canonical neighborhood in a locally compact NSS group G. Let $\{a_i\}$ be a sequence in G, and $\{m_i\}$ a sequence of positive integers. In this context we write $<a_i, m_i>$ for the sequence of pairs. We say that the sequence is <u>standard</u> if the following is true: $a_i \rightarrow 1$, $a_i \in U$, $m_i \rightarrow \infty$, and there exists a positive real number k such that

$$a_i, a_i^2, \ldots, a_i^{[km_i]} \in U$$

for all i. (Strictly speaking, we should say "standard relative to U", but U will be held fixed so much that there will be no danger of ambiguity.) We call k a <u>modulus</u> of the sequence.

<u>Definition.</u> We say that the standard sequence $<a_i, m_i>$ converges to the one-parameter subgroup X if $a_i^{[rm_i]} \rightarrow X(r)$ for all $r \geq 0$. We simply say that $<a_i, m_i>$ converges, if we do not wish to specify X.

<u>Theorem 8.</u> Let U be a canonical neighborhood in a locally compact NSS group. Let $<a_i, m_i>$ be a standard sequence. Let a neighborhood V of 1 be given. Then there exists a positive real number r_0 such that $a_i^{[rm_i]} \in V$ for all i and all r with $0 \leq r < r_0$.

<u>Proof.</u> Let k be a modulus for $<a_i, m_i>$. Set $p_i = [km_i]$. Theorem 4 is applicable, and tells us that there is a positive real num-

ber s such that $a_i^{[tp_i]} \in V$ for all i and all t with $0 \leq t \leq s$. We show that the choice $r_o = ks/2$ works for sufficiently large i; this will prove the theorem, since r_o can be further lowered to look after a finite number of cases. Take $r < r_o$. To show that $a_i^{[rm_i]} \in V$ it suffices to verify that $[rm_i] \leq [sp_i]$. As soon as $km_i \geq 2$, we have

$$km_i/2 \leq km_i - 1 \leq [km_i] = p_i \ ,$$

$$rm_i < r_o m_i = ksm_i/2 \leq sp_i \ ,$$

so that $[rm_i] \leq [sp_i]$.

Theorem 9. Let U be a canonical neighborhood in a locally compact NSS group. Let $<a_i, m_i>$ be a standard sequence with modulus k. If $a_i^{[km_i]}$ converges, then $<a_i, m_i>$ converges.

This is immediate from Theorem 5, and Theorem 10 is a corollary.

Theorem 10. Any standard sequence has a convergent subsequence.

3. Differentiable functions

We now begin the procedure that will ultimately put a Lie algebra structure on the one-parameter subgroups of a locally compact NSS group. In the last analysis this amounts to differentiating something . We choose to do our calculus in the space C of continuous real functions on G with compact support (a function has compact support if it vanishes outside a compact set, the compact set depending on the function). For the moment, G can be any topological group. We shall henceforth reserve the letter C for use in this way.

Any function in C is bounded. We norm C by setting $\|f\| = \sup_{x \in G} |f(x)|$. C is usually not complete, but that will not disturb us (of course, C is complete if G is compact).

We introduce <u>left translation</u>. Given $a \in G$ and $f \in C$ we define af by $(af)x = f(a^{-1}x)$. We have $af \in C$ (if f is supported by the compact set W, then af is supported by aW). Also, $\|af\| = \|f\|$.

Let F be a function from the real numbers to C, defined in a neighborhood of r. If

$$\lim_{h \to 0} \frac{F(r+h) - F(r)}{h}$$

exists, we say that F is differentiable at r. Given $f \in C$ and a one-parameter subgroup X we can define F by $F(r) = X(r)f$; F is thus a function from the entire real line to C. The following is an easy exercise: if F is differentiable at 0, it is differentiable everywhere; in the notation of the next definition the derivative of $X(r)f$ at r is $X(r)D_X f$.

<u>Definition.</u> $f \in C$ is <u>differentiable</u> if $D_X f$ exists for any one-parameter subgroup X.

A priori, we do not know of the existence of a single differentiable $f \neq 0$. Our program is to construct a particularly useful one when G is locally compact NSS. This will be achieved after a long discussion.

We need a continuity property of left translation. The metrizable case suffices for us; the general case can be found for example in [14, Th. 28, p. 109].

<u>Theorem 11.</u> Let C be the space of continuous real functions with compact support on a metrizable locally compact group G. Then the

function af is jointly continuous from $G \times C$ to C.

Proof. As regards the variable a, it suffices to prove continuity at $a = 1$. In view of the estimate

$$\|ag - f\| = \|ag - af + af - f\| \leq \|g - f\| + \|af - f\|,$$

it suffices to do the following: given $\varepsilon > 0$ we must find a neighborhood V of 1 such that $a \in V$ implies $\|af - f\| < \varepsilon$. Suppose f vanishes outside the compact set T. Fix a compact neighborhood W of 1, and restrict a to be in W. Then af vanishes outside the compact set WT. We have that f is uniformly continuous on WT. If ρ is the distance function, there exists δ so that $\rho(x, y) < \delta$ implies $|f(x) - f(y)| < \delta$ for $x, y \in WT$. By a typical compactness argument we can find a symmetric neighborhood V of 1, $V \subseteq W$, such that $\rho(ax, x) < \delta$ for $a \in V$, $x \in WT$. Then $|f(a^{-1}x) - f(x)| < \varepsilon$ for $a \in V$ and all x, so that $\|af - f\| < \varepsilon$ and V fulfils the requirements.

We summarize some facts about equicontinuity. A set $\{f_i\}$ of real-valued functions on a topological space X is equicontinuous at a point $x \in X$ if for any $\varepsilon > 0$ there exists a neighborhood V of x such that $|f_i(y) - f_i(x)| < \varepsilon$ for all i and all $y \in V$. Suppose that X is compact metric and that the sequence $\{f_i\}$ is uniformly bounded and equicontinuous at every point of X. Then it is known that there exists a subsequence of $\{f_i\}$ that converges uniformly.

We end this section with an easy but basic result relating standard sequences, equicontinuity and differentiability.

Theorem 12. Let U be a canonical neighborhood in a locally compact NSS group. Let $<a_i, m_i>$ be standard, converging to X. Let $\{f_i\}$ be a sequence in C such that $f_i \rightarrow f$ in the norm of C, all f_i

vanish outside a fixed compact set T, and the sequence $\{m_i(a_if_i - f_i)\}$ is uniformly bounded and equicontinuous. Then: D_Xf exists and equals $\lim m_i(a_if_i - f_i)$.

Proof. The functions $a_if_i - f_i$ all vanish outside the compact set UT. By the theorem just quoted on equicontinuity, it will therefore suffice to assume that $m_i(a_if_i - f_i) \to g$ and show that $g = D_Xf$. Given $\varepsilon > 0$ we must name $\delta > 0$ so that $0 < |h| < \delta$ implies

(1)
$$\left\| \frac{X(h)f - f}{h} - g \right\| \le \varepsilon .$$

We shall assume $h > 0$, leaving it to the reader to adjust the argument for negative h. By Theorem 11, we can pick a neighborhood V of 1 such that $a \in V$ implies $\|ag - g\| < \frac{\varepsilon}{2}$. By Theorem 8 there exists a positive δ (we can take $\delta \le 1$) such that $a_i^{[rm_i]} \in V$ for all i and all r with $0 \le r < \delta$. We claim that this δ fulfils our requirement.

Let us abbreviate $[hm_i]$ to d. The following is an identity:

(2)
$$m_i(a_i^d f_i - f_i) - dg = \sum_{t=0}^{d-1} a_i^t[m_i(a_if_i - f_i) - g] + \sum_{t=0}^{d-1} (a_i^t g - g).$$

Since $h < \delta$, we have $a_i, \ldots, a_i^{d-1} \in V$. Hence we get the estimate $\frac{d\varepsilon}{2}$ for the last term of (2). We may assume that i is so large that

$$\|m_i(a_if_i - f_i) - g\| < \frac{\varepsilon}{2} .$$

Left-translation by a_i^t preserves the norm in C. Hence we get the same estimate $\frac{d\varepsilon}{2}$ for the first term on the right of (2). Combining these estimates, and dividing (1) by d, we find

(3)
$$\|d^{-1}m_i(a_i^d f_i - f_i) - g\| < \varepsilon .$$

Let $i \to \infty$ in (3). Since $m_i \to \infty$,

$$\frac{m_i}{d} = \frac{m_i}{[hm_i]} \to \frac{1}{h} .$$

Also, $a_i^{[hm_i]} \to X(h)$ and $f_i \to f$. In the limit, with the aid of Theorem 11, we have the desired estimate (1).

4. Functions constructed from a single Q.

We isolate in this section a construction that starts from sets U, Q and turns out functions $\Delta, \emptyset, \psi$. In the next section this will be amplified to a sequence of Q's.

We assume G to be a metrizable locally compact group and U a compact symmetric neighborhood of 1. (In §4 and §5, U is not assumed to be canonical; indeed G need not be NSS.) The set Q is symmetric, contains 1, and has the following two properties: $Q \subset U$, and the subgroup generated by Q is not contained in U. From these ingredients, Δ emerges uniquely; $\emptyset$ and ψ are unique after a "smoothing kernel" θ has been selected.

We subdivide the treatment under separate headings.

(i). The integer n. We take n to be the smallest positive integer such that $Q^n \not\subset U$. (Observe that Q^n denotes not just the set of products of n-th powers of elements of Q, but the set of products of n -- possibly different -- elements of Q.) Such an integer exists since the subgroup generated by Q does not lie in U.

(ii) The function Δ. Define a function Δ on G by $\Delta(1) = 0$,

$$\Delta(x) = 1/n \quad \text{if} \quad x \in Q, x \neq 1,$$

$$\Delta(x) = 2/n \quad \text{if} \quad x \in Q^2, x \notin Q,$$

. . . .

$$\Delta(x) = (n-1)/n \text{ if } x \in Q^{n-1} , x \notin Q^{n-2}$$

$$\Delta(x) = 1 \text{ if } x \notin Q^{n-1}$$

<u>Lemma 3.</u>   (a) $\Delta = 1$ on the complement of U.   (b) $0 \le \Delta \le 1$.
(c) $\Delta(x) = \Delta(x^{-1})$.   (d) $\Delta(xy) \le \Delta(x) + \Delta(y)$.   (e) $|\Delta(qx) - \Delta(x)| \le 1/n$
for $q \in Q$.

<u>Proof.</u>   (a),(b), and (c) are obvious.

(d). If $\Delta(x)$ or $\Delta(y)$ is 1 or 0, the result is vacuous. Assume
$\Delta(x) = i/n$ and $\Delta(y) = j/n$ with $i, j < n$. Then $x \in Q^i$, $y \in Q^j$, so that
$xy \in Q^{i+j}$ and $\Delta(xy) \le (i+j)/n$.

(e). By (d) we have $\Delta(qx) \le \Delta(q) + \Delta(x)$, and $\Delta(q) = 0$ or $1/n$
since $q \in Q$. Since $x = q^{-1} \cdot qx$, we similarly find $\Delta(x) \le \Delta(qx) + 1/n$.
Hence $|\Delta(qx) - \Delta(x)| \le 1/n$.

(iii). <u>The function θ.</u>   The function Δ is a step function with
steps reflecting the "growth" of the powers of Q. We wish to have a
continuous function with similar properties, and we shall achieve this
by smoothing Δ via an appropriate continuous function θ. It will turn
out to be vital for θ, like Δ, to have a strict maximum at 1. We shall
call a real function on G <u>proper</u> if it has a strict maximum at 1.

The properties desired for θ are as follows: $\theta(1) = 1$, $0 \le \theta \le 1$,
θ is proper, and θ vanishes outside U. Using the metric on G and
the Tietze extension theorem, it is easy to construct such a function.

(iv) <u>The function ϕ.</u>   Define

(4) $\phi(x) = \sup_{y \in G} [\{1 - \Delta(y)\} \theta(y^{-1}x)].$

<u>Lemma 4.</u> (a) ϕ vanishes outside U^2. (b) $\phi(1) = 1$, $0 \le \phi \le 1$,
ϕ is proper. (c) $|\phi(qx) - \phi(x)| \le 1/n$ for $q \in Q$.

Proof. (a) The factor $1 - \Delta(y)$ in (4) vanishes unless $y \in U$. The factor $\theta(y^{-1}x)$ vanishes unless $y^{-1}x \in U$. Hence the product vanishes unless $x \in U^2$.

(b) That $0 \leq \phi \leq 1$ is clear, since both factors in (4) lie between 0 and 1. Both factors are 1 for $x = y = 1$; hence $\phi(x) = 1$. Suppose that $x \neq 1$. Now the factor $1 - \Delta(y)$ is bounded by $1 - 1/n$ for $y \neq 1$. For $y = 1$, the factor $\theta(y^{-1}x) < 1$. Hence $\phi(x) < 1$ for $x \neq 1$, so that ϕ is proper.

(c) We have

$$(5) \qquad \phi(qx) = \sup_{y \in G} \; [\{1 - \Delta(y)\} \, \theta(y^{-1}qx)].$$

Since the sup is being taken over all y, we can, in (5), replace y by qy:

$$(6) \qquad \phi(qx) = \sup_{y \in G} \; [\{1 - \Delta(qy)\} \, \theta(y^{-1}x)].$$

By part (e) of Lemma 3, $1 - \Delta(qy)$ and $1 - \Delta(y)$ differ by at most $1/n$. Hence comparison of (4) and (6) yields $|\phi(qx) - \phi(x)| \leq 1/n$.

We do not pause at this point to prove the continuity of ϕ, since in §5 we shall in fact prove the equicontinuity of a whole sequence of ϕ's.

(v) The function ψ. We pass from ϕ to ψ by a second smoothing. Here our intention, roughly speaking, is to make ψ differentiable. So it is reasonable to get ψ by integration. The moment has therefore come to invoke the existence of a left invariant Haar measure on G. The point of view that measure is a positive linear function on C (the continuous real functions with compact support) fits our needs admirably, since these are the only functions we shall integrate. We shall however make occasional reference to the measure of a compact neighborhood V of 1, writing it $m(V)$; this could be obviated by integrating a suitable element of C. We write $\int f(x)\,dx$ for the integral.

The definition of ψ is as follows:

$$\psi(u) = \int \phi(ux)\,\phi(x)\,dx.$$

At present we note only one property of ψ.

Lemma 5. ψ vanishes outside U^4.

Proof. Since ϕ vanishes outside U^2 (Lemma 4 a), the integrand is 0 unless both x and ux lie in U^2, whence $u \in U^4$.

5. Functions constructed from a sequence of Q's.

The basic setup continues that of §4, starting with a metrizable locally compact group G, and a compact symmetric neighborhood U of 1. Instead of a single Q, we have a sequence $Q_1, Q_2, \ldots$. Each Q_i is symmetric, $1 \in Q_i$, $Q_i \subset U$, and the subgroup generated by Q_i is not contained in U. The results set forth in this section will be used twice. The first time (§§6-9), G will be NSS, U canonical, and Q_i will be the set of all x such that $x, x^2, \ldots, x^i \in U$. The second time (§10) we shall take a fundamental sequence V_i of neighborhoods of 1, and define Q_i to be the set-theoretic union of the subgroups contained in V_i. The assumption that the subgroup generated by Q_i is not contained in U will on this second occasion be an indirect one, on the road to the proof of Theorem 17.

From Q_i, as in §4, we get $n_i, \Delta_i, \phi_i, \psi_i$. A single fixed θ is to be used in the passage from Δ_i to ϕ_i.

Lemma 6. Given $\varepsilon > 0$, there exists a neighborhood V of 1 such that

$$\left| \emptyset_i(xa) - \emptyset_i(x) \right| < \varepsilon$$

for all i, all $x \in G$, and all $a \in V$.

Proof. By Theorem 11, we can pick V so that $\left| \theta(xa) - \theta(x) \right| < \varepsilon$ for all $x \in G$ and all $a \in V$ (although Theorem 11 was stated and proved for left translation, it is of course equally valid for right translation). Then

$$(7) \qquad \emptyset_i(xa) - \emptyset_i(x) = \sup_y \left[\{1 - \Delta_i(y)\} \, \theta(y^{-1}xa) \right]$$
$$- \sup_y \left[\{1 - \Delta_i(y)\} \, \theta(y^{-1}x) \right].$$

For each fixed y, the terms on the right of (7) differ by at most ε. Hence the two sup's differ by at most ε.

Lemma 7. The set $\{\emptyset_i\}$ is equicontinuous.

Proof. We fix a point $x \in G$ and have to prove equicontinuity there. Given $\varepsilon > 0$, we pick xV as our neighborhood of x, where V is given by the preceding theorem.

We next handle left translation.

Lemma 8. Given $\varepsilon > 0$, there exists a neighborhood V of 1 such that

$$\left| \emptyset_i(ax) - \emptyset_i(x) \right| < \varepsilon$$

for all x, all i, all $a \in V$.

Proof. Suppose the contrary. Then there will exist a sequence $a_j \to 1$, a sequence $x_j \in G$, and a subsequence $\{\emptyset_j\}$ of the $\emptyset$'s such that

$$(8) \qquad \left| \emptyset_j(a_j x_j) - \emptyset_j(x_j) \right| \geq \varepsilon \quad .$$

By equicontinuity (Lemma 7), and the fact that the $\emptyset$'s are uniformly bounded and vanish outside U^2 (Lemma 4 a), we can suppose that $\emptyset_j \to \sigma$ in C. We can also assume $a_j \in U$. Then it follows that the x_j's must lie in U^3. We can therefore assume $x_j \to x$ (change notation, as always). We have $\emptyset_j(x_j) \to \sigma(x)$ and $\emptyset_j(a_j x_j) \to \sigma(x)$, contradicting (8).

Lemma 9. ψ_i is continuous.

Proof.   This is a routine consequence of the definition of ψ_i as
$$\psi_i(u) = \int \emptyset_i(ux)\emptyset_i(x)dx \ ,$$
and the continuity of $\emptyset_i$.

Lemma 10.   The set of all functions $n_i(q_i\psi_i - \psi_i)$, where i ranges over the positive integers and q_i ranges over Q_i, is uniformly bounded, equicontinuous, and vanishes outside U^5.

Proof.   Since $q_i \in Q_i \subset U$ and ψ_i vanishes outside U^4 (Lemma 5), the vanishing of $q_i\psi_i - \psi_i$ outside U^5 is clear.

We have
$$(9) \qquad n_i(q_i\psi_i - \psi_i)(u) = n_i \int [\emptyset_i(q_i^{-1}uy) - \emptyset_i(uy)]\emptyset_i(y) \, dy.$$
The quantity
$$\emptyset_i(q_i^{-1}uy) - \emptyset_i(uy)$$
within the brackets in (9) is bounded by $1/n_i$ by Lemma 4c. Hence (9) is dominated by $m(U^2)$, the measure of U^2. This establishes uniform boundedness. It remains to check equicontinuity.

We tackle equicontinuity at u. Take a symmetric neighborhood V of 1 as in Lemma 8, for the number $\varepsilon/m(U^7)$. With $z \in uV$ we prove
$$\left| n_i(q_i\psi_i - \psi_i)(z) - n_i(q_i\psi_i - \psi_i)(u) \right| < \varepsilon \ .$$

We write out the integral for $n_i(q_i \psi_i - \psi_i)(u)$, and then replace y by $u^{-1}y$, as permitted by left-invariance:

(10) $\qquad n_i(q_i \psi_i - \psi_i)(u) = n_i \int [\phi_i(q_i^{-1}y) - \phi_i(y)] \phi_i(u^{-1}y) \, dy$.

We are to replace u by z in (10) and estimate the absolute value of the difference. Repeating the argument given above, we observe that the quantity

$$\phi_i(q_i^{-1}y) - \phi_i(y)$$

within the brackets in (10) is bounded by $1/n_i$. It remains therefore to verify

(11) $\qquad \int |\phi_i(z^{-1}y) - \phi_i(u^{-1}y)| \, dy < \varepsilon$.

Now u and z may be confined to U^5, since we have verified that $n_i(q_i \psi_i - \psi_i)$ vanishes outside U^5. Since ϕ_i vanishes outside U^2, the integration in (11) may be confined to U^7. Since

$$z^{-1}y = (z^{-1}u)(u^{-1}y)$$

with $z^{-1}u \in V$, our choice of V shows that

$$|\phi_i(z^{-1}y) - \phi_i(u^{-1}y)| < \varepsilon/m(U^7) .$$

Lemma 10 is proved.

The ϕ_i's are equicontinuous, uniformly bounded, and have a fixed compact support. We can pick a subsequence converging in the norm of C. On this occasion, it will be wise for us to be a little more formal in our notation for a subsequence. We write $\phi_{p(i)}$ for a convergent subsequence and ϕ for the limit, and hold this notation fixed in Lemmas 11 and 12.

<u>Lemma 11.</u> $\psi_{p(i)}$ converges in C, say to ψ. We have

(12) $\qquad \psi(u) = \int \phi(uy) \phi(y) \, dy$.

Proof. Let us define ψ by (12); the integral surely exists, for $\emptyset$ enjoys the same bound and compact support as the $\emptyset_i$'s. We must prove that $\psi_{p(i)} \to \psi$. Take i so large that

(13)
$$\| \emptyset_{p(i)} - \emptyset \| < \epsilon / 2m(U^2).$$

We have

(14)
$$\psi_{p(i)}(u) - \psi(u) = \int \emptyset_{p(i)}(uy)[\emptyset_{p(i)}(y) - \emptyset(y)]dy$$
$$+ \int [\emptyset_{p(i)}(uy) - \emptyset(uy)]\emptyset(y) \ dy .$$

Using (13), we make a routine estimate on (14) to obtain $\| \psi_{p(i)} - \psi \| < \epsilon$, as required.

We recall that each $\emptyset_i$ is proper, i.e., has a strict maximum at 1 (Lemma 4b). This property may be lost in the passage to the limit, and we study what happens.

Lemma 12. $\emptyset(x) = 1$ if and only if there exists a sequence $y_i \in G$ such that $y_i \to x$ and $\Delta_{p(i)}(y_i) \to 0$.

Proof. We have that $\emptyset(x) = 1$ if and only if $\emptyset_{p(i)} \to 1$. Now

(15)
$$\emptyset_{p(i)}(x) = \sup_y [\{1 - \Delta_{p(i)}(y)\} \theta(y^{-1}x)] .$$

For $\emptyset_{p(i)}(x)$ to be close to 1, both factors on the right of (15) must be close to 1, i.e., $\Delta_{p(i)}(y)$ must be close to 0, and $\theta(y^{-1}x)$ must be close to 1, which means that y must be close to x. This enables us to choose the sequence y_i as required.

Lemma 13. The set S of elements x with $\emptyset(x) = 1$ forms a subgroup contained in U.

Proof. Let x be an element of S. By Lemma 12, we have a sequence y_i with $y_i \to x$ and $\Delta_{p(i)}(y_i) \to 0$. Since any Δ is equal to 1 outside U, we must have $y_i \in U$ for large i. Hence $x \in U$. Since each Δ is symmetric, we get $x^{-1} \in S$. Finally, given another element $\overline{x}$ in S, we must prove that $x\overline{x} \in S$. Use $\overline{y}_i$ for the sequence that goes with $\overline{x}$ ($\overline{y}_i \to \overline{x}$, $\Delta_{p(i)}(\overline{y}_i) \to 0$). We have $y_i \overline{y}_i \to x\overline{x}$. By Lemma 3d,

$$\Delta(y_i \overline{y}_i) \le \Delta(y_i) + \Delta(\overline{y}_i)$$

for any Δ. Hence $\Delta_{p(i)}(y_i \overline{y}_i) \to 0$, and $x\overline{x} \in S$ by another application of Lemma 12.

An immediate consequence of Lemma 13 is: if U contains no subgroup $\ne 1$, then $\emptyset$ is proper. From $\emptyset$ we pass to ψ by the Cauchy inequality.

Lemma 14. If $\emptyset$ is proper, so is ψ.

Proof. We apply the Cauchy inequality to formula (12), obtaining:

(16) $$\psi(u)^2 \le \int \emptyset(uy)^2 dy \int \emptyset(y)^2 \, dy \ .$$

By left invariance, the two integrals on the right side of (16) are equal, and by (12) again, they are equal to $\psi(1)$. One has equality in the Cauchy inequality (16) only if $\emptyset(uy) = \lambda\emptyset(y)$ for some constant λ. Putting $y = 1$ and $y = u^{-1}$ in succession shows that $\lambda = 1$, and then that $u = 1$.

6. Proof that i/n_i is bounded.

We are in a position to prove a key result.

Theorem 13. Let U be a canonical neighborhood in a locally compact NSS group G. Define Q_i to be the set of all x satisfying $x, x^2, \ldots, x^i \in U$. Let n_i be the smallest positive integer such that

$Q_i^{n_i} \not\subset U$. Then: i/n_i is bounded.

First we note a lemma.

Lemma 15. The Q_i's form a decreasing fundamental sequence of compact neighborhoods of 1. Also $n_i \to \infty$.

Proof. That the Q_i's form a decreasing sequence of compact neighborhoods of 1 is obvious. The "fundamental" property means that any neighborhood W of 1 contains one of the Q_i's. If we deny this, we get a sequence $\{z_i\}$ with $z_i \in Q_i$, $z_i \notin W$. Let z be a limit point of $\{z_i\}$; necessarily $z \neq 1$. Then any power z^p is a limit point of $\{z_i^p\}$. Since $z_i^p \in U$ for $i \geq p$, we have $z^p \in U$. Thus the whole subgroup generated by z lies in U, a contradiction. That $n_i \to \infty$ is an immediate consequence.

Proof of Theorem 13. We invoke the machinery of §5, leading up to ϕ_i and ψ_i. Pick, for each i, any element c_i lying in $Q_i^{n(i)}$ but not in U. We have that c_i is a product of n_i elements, each in Q_i. Select one of these, say a_i, maximizing $\|a_i\psi_i - \psi_i\|$ among these n_i terms. By repeated use of the estimate

$$\|abf - f\| \leq \|af - f\| + \|bf - f\|,$$

we get

(17) $$\|c_i\psi_i - \psi_i\| \leq n_i\|a_i\psi_i - \psi_i\|.$$

We make an indirect proof, assuming that i/n_i is unbounded. In choosing, as in §5, a sequence $\{p(i)\}$ of integers such that $\phi_{p(i)} \to \phi$, we can then furthermore arrange that $p(i)/n_{p(i)} \to \infty$.

Observe that, for every i, $a_i, a_i^2, \ldots, a_i^i$ all lie in U since $a_i \in Q_i$, and recall (Lemma 15) that $a_i \to 1$ and $n_i \to \infty$. Of course, $n_i \leq i$. In the terminology introduced in §2, $\langle a_i, i \rangle$ and $\langle a_i, n_i \rangle$ are standard (the modulus can be taken as 1 for both). We note the result of applying Theorem 4 or Theorem 8 to $\langle a_i, i \rangle$: for any neighborhood V of 1 there exists $r_o > 0$ such that $a_i^{[si]} \in V$ for all i and all $s < r_o$.

We turn to the sequence $\langle a_i, n_i \rangle$ and drop to the subsequence $\langle a_{p(i)}, n_{p(i)} \rangle$. We claim that this converges to the trivial one-parameter subgroup 0 (we write 0 for the one-parameter subgroup X defined by $X(r) = 1$ for all r). For let V be any neighborhood of 1. Fix $r > 0$. With r_o chosen as in the last paragraph, we have

$$\frac{rn_{p(i)}}{p(i)} < r_o$$

for large i, since $n_{p(i)}/p(i) \to 0$. Hence

$$a_{p(i)}^{[\frac{rn_{p(i)}}{p(i)} p(i)]} \in V$$

i.e.,

$$a_{p(i)}^{[rn_{p(i)}]} \in V.$$

Therefore

$$a_{p(i)}^{[rn_{p(i)}]} \to 1.$$

This is true for any r. We have shown that the standard sequence $\langle a_{p(i)}, n_{p(i)} \rangle$ converges to the one-parameter subgroup 0.

By Lemma 10, the sequence $\{n_i(a_i\psi_i - \psi_i)\}$ is uniformly bounded, equicontinuous, and vanishes outside U^5. Of course, the same is true for the subsequence $\{n_{p(i)}(a_{p(i)}\psi_{p(i)} - \psi_{p(i)})\}$. Furthermore, $\psi_{p(i)} \to \psi$ by Lemma 11. This puts us in a position to apply Theorem 12. The conclusion is that $n_{p(i)}(a_{p(i)}\psi_{p(i)} - \psi_{p(i)})$ converges to $D_0\psi$, and $D_0\psi$ is, of course, 0. From (17) we next deduce that

$$c_{p(i)}\psi_{p(i)} - \psi_{p(i)} \to 0 \ .$$

If c is a limit point of $\{c_{p(i)}\}$, we have $c\psi = \psi$. But ψ is proper (Lemmas 13 and 14). Hence $c = 1$, a contradiction since the elements $\{c_{p(i)}\}$ lie outside U. This concludes the proof of Theorem 13.

7. Existence of proper differentiable functions.

In this section G is a locally compact NSS group, U is a canonical neighborhood, and we continue with the full notation of §5.

We first apply Theorem 13 to improve Lemma 10 by replacing n_i by i; this is clearly legal since Theorem 13 asserts that i/n_i is bounded. For explicitness, we state the revised version.

Lemma 16.   The set of all functions $i(q_i\psi_i - \psi_i)$, where i ranges over the positive integers and q_i ranges over Q_i, is uniformly bounded, equicontinuous, and vanishes outside U^5.

It will be convenient to introduce the notation Ψ for the set of all functions ψ constructed at the end of §5. We repeat the relevant details: from the sequence $\{\emptyset_i\}$ a convergent subsequence $\{\emptyset_{p(i)}\}$ was extracted, converging to $\emptyset$; the corresponding subsequence $\psi_{p(i)}$ converged to ψ, a typical member of Ψ.

Lemma 17. Any $\psi \in \Psi$ is a proper differentiable function.

Lemma 18. Let K_1 be the set of all one-parameter subgroups X such that $X(r) \in U$ for $|r| \leq 1$. Then for any $\psi \in \Psi$ the set $\{D_X\psi\}$, X ranging over K_1, is uniformly bounded, equicontinuous, and vanishes outside a fixed compact set.

Proof. We prove Lemmas 17 and 18 together. We note at once that any ψ in Ψ is proper by Lemmas 13 and 14.

We introduce at this point the process of changing the parameter in a one-parameter subgroup X. For λ a real number we define λX by $(\lambda X)(r) = X(\lambda r)$. It is routine to check that λX is again a one-parameter subgroup, and the following observation is likewise routine: for any $f \in C$, $D_{\lambda X}f$ exists if $D_X f$ exists, and in that case $D_{\lambda X}f = \lambda D_X f$. Because of this observation we can , in Lemma 17, confine ourselves to one-parameter subgroups lying in K_1.

Let $X \in K_1$ and set $a_i = X(1/i)$. Then $<a_i, i>$ is a standard sequence converging to X. Also, $a_i \in Q_i$. By Lemma 16, $\{i(a_i\psi_i - \psi_i)\}$ is uniformly bounded, equicontinuous, and vanishes outside U^5. Take any $\psi \in \Psi$, and the subsequence $\psi_{p(i)} \to \psi$. Theorem 12 is applicable and shows that $D_X\psi$ exists and equals the limit of the sequence

$$\{p(i)(a_{p(i)}\psi_{p(i)} - \psi_{p(i)})\} \ .$$

This proves Lemma 17. Now let X range over K_1. The properties of uniform boundedness, equicontinuity and possession of a fixed compact

114

support are preserved under the passage to uniform limits. The statements in Lemma 18 follow.

8. The vector space of one-parameter subgroups.

Throughout this section G is again a locally compact NSS group with a fixed canonical neighborhood U. We use the letter f to denote a fixed proper differentiable function which further has the property asserted for ψ in Lemma 18; we switch the notation from ψ to f in order to emphasize that in this section the source of f does not matter.

Lemma 19. If $< a_i, m_i >$ and $< b_i, m_i >$ are standard, so is $< a_i b_i, m_i >$. Suppose further that $< a_i, m_i >$ converges to X, $< b_i, m_i >$ to Y, and $< a_i b_i, m_i >$ to Z, and that $m_i(a_i f - f) \to D_X f$, $m_i(b_i f - f) \to D_Y f$. Then $m_i(a_i b_i f - f) \to D_Z f$ and $D_Z f = D_X f + D_Y f$.

Proof. We can assume the same modulus k for both $< a_i, m_i >$ and $< b_i, m_i >$ (take the smaller one). Pick a positive integer A which bounds i/n_i (Theorem 13). We have

$$a_i, b_i \in Q_{[km_i]} \quad .$$

Hence $(a_i b_i)^s \in U$ if $2s \leq n_{[km_i]}$. Since $n_i/i \geq 1/A$, we have

$$(18) \qquad n_{[km_i]} \geq \frac{[km_i]}{A} \quad .$$

We claim that $< a_i b_i, m_i >$ is standard with modulus $k/2A$. To see this we have to verify that the powers of $a_i b_i$ lie in U, up to the power $[km_i/2A]$. This will be so provided

$$[\frac{km_i}{2A}] \leq \frac{n_{[km_i]}}{2} \quad .$$

In the light of (18), what we need is

$$\left[\frac{km_i}{2A}\right] \leq \frac{[km_i]}{2A} \ .$$

Now $2A[km_i/2A]$ is an integer $\leq km_i$, and therefore $\leq [km_i]$, as required.

We turn to the proof of the second part of the lemma. We have

$$m_i(a_i b_i f - f) = a_i\{m_i(b_i f - f)\} + m_i(a_i f - f).$$

From our hypothesis, the fact that $a_i \to 1$, and Theorem 11, we deduce that

$$m_i(a_i b_i f - f) \to D_Y f + D_X f.$$

Now a convergent sequence is certainly uniformly bounded and equicontinuous. Moreover $\{m_i(a_i b_i f - f)\}$ has a fixed compact support. Theorem 12 is applicable to tell us that $m_i(a_i b_i f - f) \to D_Z f$. Thus all the statements in Lemma 19 stand proved.

We proceed to a series of arguments that will ultimately put the structure of a vector space on the one-parameter subgroups. We already have a zero element, and multiplication by real scalars was introduced during the proof of Lemma 17.

Definition. Let X, Y, Z be one-parameter subgroups. We say that $X + Y$ exists and equals Z if, for every $r \geq 0$,

$$\{X(1/i)Y(1/i)\}^{[ri]} \to Z(r).$$

Lemma 20. If $X + Y$ exists and equals Z, then $D_Z f = D_X f + D_Y f$.

Proof. Observe that $<X(1/i), i>$ is standard, and that

$$i\{X(1/i)f - f\} \to D_X f$$

since f is differentiable. It remains to quote Lemma 19, with a_i, b_i, m_i replaced by $X(1/i), Y(1/i), i$.

We insert at this point a quite easy result.

Lemma 21. $D_X f = 0$ implies $X = 0$.

Proof. We begin with a remark on a function F from the real numbers to C: if $F'(r)$ exists and equals 0 for all r, then F is a constant. To see this, pick $u \in G$ and write $g(r) = (F(r))(u)$. Then $g'(r) = 0$ for all r, so g is a constant, and F is a constant.

We apply this to $F(r) = X(r)f$. We have $F'(r) = X(r)D_X f = 0$, whence F is a constant. So $X(r)f = X(0)f = f$. Recalling our running assumption that f is proper, we conclude that $X(r) = 1$ for all r.

Lemma 22. $X + Y$ exists and equals 0 if and only if $D_X f + D_Y f = 0$.

Proof. The "only if" part is a special case of Lemma 20.

Suppose that $D_X f + D_Y f = 0$. By Lemma 19, $<X(1/i)Y(1/i), i>$ is standard. Consider a convergent subsequence of $<X(1/i)Y(1/i), i>$, converging say to Z. By Lemma 19, $D_Z f = D_X f + D_Y f$ and this is 0 by hypothesis. By Lemma 21, $Z = 0$. So every convergent subsequence of $<X(1/i)Y(1/i), i>$ converges to 0. By Theorem 10 we can assert that every subsequence has a subsequence converging to 0. Hence the whole sequence converges to 0, that is, $X + Y$ exists and equals 0.

Lemma 23. If $X + Y$ exists and equals 0, then $X = -Y$.

Proof. Fix for the moment a real number s. Apply to G the inner automorphism, say σ, induced by $Y(s)$; σ leaves Y fixed and sends X into a conjugate one-parameter subgroup X_1. We have by hypothesis that $X + Y$ exists and equals 0. By applying σ we deduce the statement that $X_1 + Y$ exists and equals 0. By Lemma 22, $D_X f + D_Y f = 0$ and $D_{X_1} f + D_Y f = 0$, so that $D_{X_1} f = D_X f = -D_Y f$.

Define $F(r, s) = X(r)Y(s)f$. Noting that $X(r)Y(s) = Y(s)X_1(r)$, we get

$$\frac{\partial F(r, s)}{\partial r} = \frac{\partial}{\partial r}\left[Y(s)X_1(r)f\right] = Y(s)X_1(r)D_{X_1}f$$

$$= -X(r)Y(s)D_Y f = \frac{-\partial F(r, s)}{\partial s}.$$

Hence $F(r, s)$ is a function of $r - s$ (this familiar fact for numerical functions is available for the vector-valued function F by evaluating F at a group element). In particular, $F(r, 0) = F(0, -r)$ and $X(r)f = Y(-r)f$. We deduce $X(r) = Y(-r)$ since f is proper, i.e. $X = -Y$ as required.

Lemma 24. $D_X f = D_Y f$ implies $X = Y$.

Proof. From $D_X f = D_Y f$ we get, by Lemma 22, that $X + (-Y)$ exists and equals 0. By Lemma 23, $X = -(-Y) = Y$.

We are at length ready to prove that $X + Y$ exists.

Lemma 25. For any X and Y, $X + Y$ exists. We have $D_{X+Y}f = D_X f + D_Y f$.

Proof. We argue as in Lemma 22. If a subsequence of $< X(1/i)Y(1/i), i >$ converges to Z, then $D_Z f = D_X f + D_Y f$ by Lemma 19. Lemma 24 now shows that all such Z's are the same. Hence $X + Y$ exists.

We introduce the notation L (or $L(G)$ if it is advisable to call attention to G) for the set of one-parameter subgroups in G. The following lemma summarizes information we have already acquired.

Lemma 26. The map $L \to C$ given by $X \to D_X f$ is one-to-one, preserves addition, and preserves scalar multiplication. Hence L is a vector space over the field of real numbers, relative to the addition and scalar multiplication we have introduced.

We recall some earlier notation: K_1 is the set of X with $X(t) \in U$ for $|t| \leq 1$, and K is the set of $X(1)$ for $X \in K_1$. We have the "exponential" map $X \to X(1)$ from L to G. Uniqueness of the square root shows that this map is one-to-one when restricted to K_1, so there is an inverse from K to K_1. We compound this with the map $X \to D_X f$ from K_1 to C and get $x \to D_X f$ from K to C.

Lemma 27.   The map just defined from K to C is continuous.

Proof.   In proving this we invoke our assumption that f has the property asserted for ψ in Lemma 18.

Given $x_i \to x$ in K, let $x_i = X_i(1)$, $x = X(1)$ with X_i and X in K_1. We must prove that $D_{X_i} f \to D_X f$. In view of the assumption on f just mentioned, we may assume that $D_{X_i} f \to g$, and our problem is to prove that $g = D_X f$. If we fix i for the moment, we can, by the very definition of derivative, pick m_i so large that $m_i \{ X_i(1/m_i)f - f \}$ is as close as we please to $D_{X_i} f$. Then

$$m_i \{ X_i(1/m_i)f - f \} \to g.$$

The sequence $< X_i(1/m_i), m_i >$ is standard, converging to X. We apply Theorem 12, with all the f_i's of that theorem taken to be f. The needed equicontinuity and uniform boundedness of the sequence $m_i \{ X_i(1/m_i)f - f \}$ is assured since it is convergent, and the fixed compact support is clear. Hence

$$m_i \{ X_i(1/m_i)f - f \} \to D_X f$$

and $g = D_X f$ follows.

The next lemma continues the study of the map from K to C.
Write M for the subspace of C consisting of all $D_X f$, $X \in L$.

Lemma 28.   The above map from K to C fills a neighborhood of
0 in M.

Proof.   If not, there exists a sequence $X_i \in L$ with $X_i \notin K_1$, but
$D_{X_i} f \to 0$. There exists, for each i, a number $\lambda_i < 1$ such that
$X_i(\lambda_i)$ is on the boundary of U and $X_i(t) \in U$ for $|t| \leq \lambda_i$. Write
$x_i = X_i(\lambda_i)$, and note that $x_i \in K$, $\lambda_i X_i \in K_1$. Since K is compact
(Theorem 7) we can assume that x_i converges to an element $x \in K$.
Observe that $x \neq 1$ since the x_i's are on the boundary of U. Let
$X \in K_1$ correspond to x. By Lemma 27, $D_{\lambda_i X_i} f \to D_X f$. Since
$D_{\lambda_i X_i} f = \lambda_i D_{X_i} f$, $\lambda_i < 1$, and $D_{X_i} f \to 0$, we have $D_{\lambda_i X_i} f \to 0$.   Hence
$D_X f = 0$. By Lemma 21, $X = 0$, $x = 1$, a contradiction which proves
the lemma.

Now the image of K in M is compact (Lemma 27). This makes
M locally compact. By a well-known theorem, a locally compact
normed linear space is finite-dimensional. We summarize:

Theorem 14.   In any locally compact NSS group, the one-parameter subgroups form a finite-dimensional vector space over the field
of real numbers, the vector space operations being those defined above.

9. Proof that K is a neighborhood of 1.

This section is devoted to proving the statement in the title. We continue to assume that G is a locally compact NSS group with canonical neighborhood U.

We first have to prove a result generalizing some of the work in the preceding section to standard sequences.

Lemma 29.   Let $<a_i, m_i>$ be standard, converging to X, and $<b_i, m_i>$ standard, converging to Y. Then $<a_i b_i, m_i>$ is standard, converging to X + Y.

Proof.   That $<a_i b_i, m_i>$ is standard was proved in Lemma 19. We may assume that k is a common modulus for all three sequences. Since any standard sequence has a convergent subsequence (Theorem 10), it will suffice for us to assume that $<a_i b_i, m_i>$ converges to Z and prove that Z = X + Y.

Write $h_i = [km_i]$, for brevity. We have $a_i \in Q_{h_i}$, and similarly for b_i and $a_i b_i$. We now make a further application of the technique and notation of §§5-7. The functions $\{\phi_{h_i}\}$ form a subsequence of $\{\phi_i\}$. We take the subsequence $\{\phi_{p(i)}\}$ of § 5 to be a subsequence of $\{\phi_{h_i}\}$. The functions $h_i(a_i \psi_{h_i} - \psi_{h_i})$ are, by Lemma 16, uniformly bounded, equicontinuous and vanish outside a fixed compact set. Since $h_i/m_i \to k$ as $i \to \infty$, we can assert the same for the functions $m_i(a_i \psi_{h_i} - \psi_{h_i})$. When we drop to the subsequence corresponding to $\phi_{p(i)}$ (we do not attempt to write it down, as the notation is getting complex) we have a setup to which Theorem 12 is applicable. With ψ the limit of $\psi_{p(i)}$, we write the result as follows:

subsequence of $m_i(a_i\psi_{h_i} - \psi_{h_i}) \to D_X\psi$.

The discussion is identical for b_i and a_ib_i, and it is (so to speak) the same subsequence:

$$\text{subsequence of } m_i(b_i\psi_{h_i} - \psi_{h_i}) \to D_Y\psi ,$$

$$\text{subsequence of } m_i(a_ib_i\psi_{h_i} - \psi_{h_i}) \to D_Z\psi .$$

In the equation

$$m_i(a_ib_i\psi_{h_i} - \psi_{h_i}) = m_ia_i(b_i\psi_{h_i} - \psi_{h_i}) + m_i(a_i\psi_{h_i} - \psi_{h_i})$$

we pass to the limit, and get $D_Z\psi = D_X\psi + D_Y\psi$. Since, in the notation of §7, we have $\psi \epsilon \Psi$, we conclude from Lemma 26 that $Z = X + Y$. This concludes the proof of Lemma 29.

We collect a number of further technical lemmas. Since our business with ψ has concluded, we revert to the notation where f is any fixed proper differentiable function.

Lemma 30. Let $a_i \epsilon G$ and positive integers m_i be given with $a_i \to 1$ and $m_i \to \infty$. Suppose that $m_i(a_if - f) \to 0$. Then $<a_i, m_i>$ is standard and converges to 0.

Proof. We claim that it is sufficient to prove that $a_i, \ldots, a_i^{m_i} \epsilon U$ for large i. For then $<a_i, m_i>$ is standard. We can assume $<a_i, m_i> \to X$ and then show that $X = 0$. Theorem 12 is available to see that

$$m_i(a_if - f) \to D_Xf.$$

Thus $D_Xf = 0$ and $X = 0$ by Lemma 21.

Suppose that it is not the case that $a_i, \ldots, a_i^{m_i} \epsilon U$ for large i. Then we can pick a subsequence, which we informally write a_j, and a sequence of positive integers t_j $(1 \leq t_j \leq m_j)$, such that $a_j^{t_j} \notin U$, $a_j^{t_j-1} \in U$. By dropping to a subsequence (and, as always, changing notation) we may assume that $a_j^{t_j-1} \to a \neq 1$. We have $a_j^{t_j} \to a$. In view of the estimate

$$\| a_j^{t_j} f - f \| \leq t_j \| a_j f - f \|,$$

and the hypothesis that $m_j(a_j f - f) \to 0$, we deduce $af - f = 0$. Since f is proper, we derive the contradiction $a = 1$.

Lemma 31. Let $a_i \in G$ and positive integers m_i be given with $a_i \to 1$ and $m_i \to \infty$. Suppose that $m_i(a_i f - f) \to D_X f$ for a suitable one-parameter subgroup X. Then: $< a_i, m_i >$ is standard and converges to X.

Proof. Let $b_i = X(-1/m_i)$. Then $< b_i, m_i >$ is standard and converges to $-X$. We have

(19) $$m_i(a_i b_i f - f) = m_i a_i(b_i f - f) + m_i(a_i f - f).$$

The right side of (19) converges to $-D_X f + D_X f = 0$. By Lemma 30, $< a_i b_i, m_i >$ is standard and converges to 0. Also, $< b_i^{-1}, m_i >$ is standard and converges to X. By Lemma 29, $< a_i, m_i >$ is standard and converges to X.

We strengthen Theorem 11 by getting uniformity on compact subsets of C.

Lemma 32. Let C_o be a compact subset of C and let a positive ε be given. Then there exists a neighborhood V of 1 in G such that $\| ag - g \| < \varepsilon$ for all $a \in V$, $g \in C_o$.

Proof. Fix $g \in C_o$ for the moment. We have $1g - g = 0$. By Theorem 11 there exist neighborhoods V_g of 1 and W_g of g such that

$\|ah - h\| < \varepsilon$ for all $a \in V_g$, $h \in W_g$. A finite number of W_g's cover C_o. We take V to be the intersection of the corresponding V_g's.

Recall that L is the vector space of one-parameter subgroups in G and that we know L be be finite-dimensional (Theorem 14). We topologize L as a Euclidean space. K_1 is the set of X with $X(r) \in U$ for $|r| \leq 1$. It follows from Lemma 28 that K_1 is a compact neighborhood of 0 in L.

Lemma 33. Let B be a compact subset of L. Let V be a symmetric neighborhood of 1 in G. There exists a positive r such that $X(t) \in V$ for all $|t| \leq r$ and all $X \in B$.

Proof. There exists a positive real number s such that $sB \subset K_1$. Take i so large that $Q_i \subset V$ (Lemma 15). Then the choice $r = s/i$ will do. For consider $X(t)$ with $X \in B$ and $|t| \leq r$. We have $(ir)X = sX \subset K_1$. Hence $X(t), X(t)^2, \ldots, X(t)^i \in U$, $X(t) \in Q_i$, so that $X(t) \in V$ as required.

The next lemma establishes a kind of uniform differentiability.

Lemma 34. Let B be a compact subset of L, and let a positive ε be given. There exists a positive number r such that

(20)
$$\left\| \frac{X(h)f - f}{h} - D_X f \right\| < \varepsilon$$

for all $X \in B$ and all h, $0 < h \leq r$.

Proof. Let C_o be the set of all $D_X f$, X ranging over B. The map $X \to D_X f$ is a linear transformation, and so C_o is compact. Let V be the neighborhood of 1 provided by Lemma 32 for the set C_o and the number $\varepsilon/2$; we can take V symmetric. Pick r to be the number provided by Lemma 33 for B and V. We take $X \in B$ and $0 < h \leq r$ and have to verify (20). Write $h = ns$ with n a positive integer and s so small that

$$\left\| \frac{X(s)f - f}{s} - D_X f \right\| < \frac{\varepsilon}{2} \quad .$$

We have

(21)
$$\frac{X(h)f - f}{h} - D_X f = \sum_{k=0}^{n-1} \frac{X(ks)D_X f - D_X f}{n}$$

$$+ \sum_{k=0}^{n-1} \frac{1}{n} X(ks) \left\{ \frac{X(s)f - f}{s} - D_X f \right\} \quad .$$

We have arranged that both terms on the right of (21) are dominated by $\varepsilon/2$. This gives us (20) as a consequence.

The next lemma asserts that a standard sequence converges uniformly.

Lemma 35. Let $<a_i, m_i>$ be standard, converging to X. Let V be a neighborhood of 1. Let $R > 0$ be a given number. Then there exists i_o such that

$$X(r)^{-1} a_i^{[rm_i]} \in V_,$$

for all $i \geq i_o$ and all r with $0 \leq r \leq R$.

Proof. Assume the contrary. Then in our usual informal subsequence notation we have numbers $r_j \leq R$ such that

(22)
$$X(r_j)^{-1} a_j^{[r_j m_j]} \notin V.$$

We can assume $r_j \to r$. Take a neighborhood W of 1 with $W^5 \subset V$. If $r > 0$, take $s < r$ so close to r that

(23)
$$a_j^{[(r_j - s)m_j]} \in W$$

for all large j (Theorem 8), and also

(24)
$$X(r)^{-1} X(s) \in W.$$

If $r = 0$, take $s = 0$. Since $<a_j, m_j>$ converges to X, we have

(25)
$$X(s)^{-1} a_j^{[sm_j]} \in W$$

for large j. Likewise, for large j,

(26)
$$X(r_j)^{-1} X(r) \in W .$$

Multiplying (26), (24), (25), and (23), we obtain

$$X(r_j)^{-1} a_j^{[sm_j]} a_j^{[(r_j - s)m_j]} \in W^4 .$$

Now

$$a_j^{[sm_j]} a_j^{[(r_j - s)m_j]} \quad \text{and} \quad a_j^{[r_j m_j]}$$

differ by a factor of a_j or a_j^2; this too lies in W for large j. Hence

$$X(r_j)^{-1} a_j^{[r_j m_j]} \in W^5 \subset V,$$

contradicting (22).

Consider an element $a \in G$, fixed for the moment. The map $X \to aXa^{-1}$ is a linear transformation of L which we denote by S_a. The set of linear transformations on L is of course given the Euclidean topology.

<u>Lemma 36.</u> The map $a \to S_a$ is continuous.

<u>Proof.</u> This being a homomorphism between topological groups, it suffices to prove continuity at 1. Given $a_i \to 1$, we must show that $S_{a_i} \to I$, where I is the identity linear transformation. This means that for a given fixed X we must prove that $S_{a_i} X \to X$. Let V be a symmetric neighborhood of 1 satisfying $V^3 \subset U$. We can assume, by an adjustment of parameter, that $X(r) \in V$ for $|r| \leq 1$, and we can also assume that $a_i \in V$. Then $a_i X a_i^{-1} \in K_1$ for all i. Since the mapping $X \to X(1)$ from K_1 onto K is a homeomorphism (Lemma 24), the

problem is reduced to verifying that $a_i X(1) a_i^{-1} \to X(1)$, which is obvious.

In the next lemma we make use of a vector-valued integral. It can be taken in Riemann's sense. Thus with F a continuous function from $[0, 1]$ to C, we define

$$\int_0^1 F(t)\, dt = \lim_{m \to \infty} \frac{1}{m} \sum_{r=0}^{m-1} F\left(\frac{r}{m}\right) .$$

If the range of F has a fixed compact support, the usual elementary discussion shows that the integral exists.

Lemma 37. Let X, Y be given and write $Z = X + Y$, $X_t = Y(-t)XY(t)$. Then

$$Z(1)f - Y(1)f = \int_0^1 Z(1-t)Y(t)D_{X_t} f\, dt.$$

Proof. The following is an identity:

$$(27) \quad \left\{Y\left(\frac{1}{m}\right)X\left(\frac{1}{m}\right)\right\}^m f - Y(1)f =$$

$$\frac{1}{m} \sum_{r=0}^{m-1} \left\{Y\left(\frac{1}{m}\right)X\left(\frac{1}{m}\right)\right\}^{m-r} X\left(-\frac{1}{m}\right)Y\left(\frac{r}{m}\right)D_{X_{r/m}} f$$

$$+ \frac{1}{m} \sum_{r=0}^{m-1} \left\{Y\left(\frac{1}{m}\right)X\left(\frac{1}{m}\right)\right\}^{m-r} X\left(-\frac{1}{m}\right)Y\left(\frac{r}{m}\right)\left\{m[X_{r/m}\left(\frac{1}{m}\right)f - f] - D_{X_{r/m}} f\right\}$$

The left side of (27) approaches $Z(1)f - Y(1)f$ as $m \to \infty$. Write the right side as $A + B$. Our plan is to show that $B \to 0$ and that A approaches the integral in the theorem.

The map $t \to X_t$ is continuous by Lemma 36. Hence the set of X_t with $0 \le t \le 1$ is compact. We are in a position to use uniform differentiability (Lemma 34), and it tells us that $B \to 0$ as $m \to \infty$.

It is routine to see that the integrand $Z(1-t)Y(t)D_{X_t} f$ enjoys a fixed compact support for $0 \leq t \leq 1$, and that is is a continuous function of t. Hence the integral exists.

We must now compare A with the Riemann approximating sum

$$A_o = \frac{1}{m} \sum_{r=0}^{m-1} Z(1 - \frac{r}{m}) Y(\frac{1}{m}) D_{X_{r/m}} f$$

and argue that $A - A_o \to 0$ as $m \to \infty$. Write C_o for the set of all $Y(t) D_{X_t} f$, $0 \leq t \leq 1$. C_o is compact. Given $\varepsilon > 0$, by Lemma 32 we can find a neighborhood V of 1 such that $\| ag - g \| < \varepsilon$ for all $a \in V$, $g \in C_o$. Take a neighborhood W of 1 with $W^2 \subset V$. We apply Lemma 35 to the standard sequence $< Y(1/m) X(1/m), m >$, which converges to Z. Using $1 - (r/m)$ for the r in Lemma 35, we obtain that

$$E = Z(1 - \frac{r}{m})^{-1} \{ Y(\frac{1}{m}) X(\frac{1}{m}) \}^{m-r} \quad \epsilon \quad W$$

for all positive integers $r \leq m$ and all sufficiently large m. Of course, $X(-1/m) \in W$ for large m. So $EX(-1/m) \in W^2 \subset V$ for large m, and hence

$$\| EX(-1/m)g - g \| < \varepsilon$$

for large m and $g \in C_o$. We take

$$g = Y(\frac{r}{m}) D_{X_{r/m}} f ,$$

and can rewrite the result as

$$\| Y(\frac{1}{m}) X(\frac{1}{m})^{m-r} X(-\frac{1}{m}) Y(\frac{r}{m}) D_{X_{r/m}} f$$

$$- Z(1 - \frac{r}{m}) Y(\frac{r}{m}) D_{X_{r/m}} f \| < \varepsilon .$$

Add over $r = 0, \ldots, m-1$ and divide by m; we get $\| A - A_o \| < \varepsilon$. Lemma 37 is proved.

Lemma 38 is a sort of variant of Lemma 33.

<u>Lemma 38.</u>   Given a neighborhood V of 1 in G, there exists a neighborhood W of 0 in L such that $X(t) \epsilon V$ for all $|t| \leq 1$ and all $X \epsilon W$.

<u>Proof.</u>   Apply Lemma 33 with $B = K_1$. If r is the positive number that Lemma 33 yields, we can take $W = rK_1$.

<u>Lemma 39.</u>   Given one-parameter subgroups X, Y_i and positive integers $N_i \to \infty$, let $Z_i = Y_i + X/N_i$. Then $< Y_i(-1)Z_i(1), N_i >$ is standard and converges to X.

<u>Proof.</u>   By Lemma 31 it suffices to prove that

(28) $$N_i[Y_i(-1)Z_i(1)f - f] \to D_X f.$$

Using Lemma 37, we rewrite the left side of (28) as

(29) $$\int_0^1 Y_i(-1)Z_i(1-t)Y_i(t)D_{X_{t,i}} f \, dt \, ,$$

where

$$X_{t,i} = Y_i(-t)X_i Y_i(t).$$

(Observe that two N_i's cancelled in the transition from (28) to (29).) It will suffice to prove the following: given $\varepsilon > 0$ we can find a positive integer i_0 such that

$$\| Y_i(-1)Z_i(1-t)Y_i(t)D_{X_{t,i}} f - D_X f \| < \varepsilon$$

for all $i \geq i_0$ and all t, $0 \leq t \leq 1$. Since the map $X \to D_X f$ is continuous (indeed linear) we can find a neighborhood W of X in L such that

$$\| D_{X_1} f - D_X f \| < \varepsilon/2$$

for $X_1 \epsilon W$. Since $a \to a^{-1}Xa$ is continuous (Lemma 36), there exists a neighborhood V of 1 such that $a^{-1}Xa \epsilon W$ for $a \epsilon V$. By Lemma

38, $Y_i(t) \in V$ for $0 \le t \le 1$ for sufficiently large i. Hence $X_{t,i} \in W$

for $0 \le t \le 1$ and sufficiently large i. Take a neighborhood V_1 of 1

such that $\| aD_X f - D_X f \| < \varepsilon/2$ for $a \in V_1$. By another application of

Lemma 38,

$$b_i = Y_i(-1)Z_i(1-t)Y_i(t)$$

lies in V_1 for $0 \le t \le 1$ and sufficiently large i. Then the estimate

$$\| b_i(D_{X_{t,i}} f - D_X f) + (b_i D_X f - D_X f) \| < \varepsilon$$

completes the proof.

<u>Lemma 40.</u> K is a neighborhood of 1.

<u>Proof.</u> If not, there exists a sequence $a_i \to 1$, $a_i \notin K$. Let us

write, for a general element $a \in G$, $N(a)$ for the integer n such that

$a, \dots, a^n \in U$ and $a^{n+1} \notin U$. By Lemma 15, $N(a) < \infty$ for $a \ne 1$. Take

$Y_i \in K_1$ such that $N_i = N(Y_i(-1)a_i)$ is as large as possible. Each N_i

is finite, for otherwise a_i would be a limit of elements in K (and

therefore lie in K since K is closed by Theorem 7). Write

$x_i = Y_i(-1)a_i$. Note that $N_i \to \infty$ since $N(x_i)$ is at least as large as

$N(a_i)$, and $N(a_i) \to \infty$ since $a_i \to 1$. Thus $x_i \to 1$, and $Y_i(-1) \to 1$,

whence $Y_i \to 0$. The sequence $<x_i, N_i>$ is standard, since

$x_i, \dots, x_i^{N_i} \in U$. After dropping to a subsequence and changing

notation, we may assume that $<x_i, N_i>$ converges, say to X. Write

$Z_i = Y_i + X/N_i$. We have $Z_i \to 0$ and so we may assume $Z_i \in K_1$.

Write $p_i = Y_i(-1)Z_i(1)$. By Lemma 39, $<p_i, N_i>$ is standard and con-

verges to X. Write $q_i = p_i^{-1}x_i$. By Lemma 29, $<q_i, N_i>$ is standard,

converging to 0. Note that

$$q_i = Z_i(-1)Y_i(1)Y_i(-1)a_i = Z_i(-1)a_i .$$

By Lemma 35 (with R taken to be 2, and V taken to be U) we have $q_i^{[rN_i]} \epsilon U$ for $0 \leq r \leq 2$ and sufficiently large i. That is,

$$q_i, \dots, q_i^{2N_i} \epsilon U .$$

This is a contradiction, for the choice of Z_i exceeds the maximum value of N_i that Y_i allegedly achieved.

Theorem 15. A locally compact NSS group is locally Euclidean.

This is a corollary of Lemma 40.

We can at this point begin to get information on homomorphic images.

Theorem 16. Let G be a locally compact NSS group and H a closed normal subgroup of G. Assume that G/H is NSS (this will ultimately be redundant). Then the natural map $L(G) \rightarrow L(G/H)$ is onto, i.e., every one-parameter subgroup of G/H can be lifted to G.

Proof. Let U_1 be a canonical neighborhood of 1 in G/H. We can pick a canonical neighborhood U of 1 in G satisfying $\pi(U) \subset U_1$, where π is the natural homomorphism from G to G/H. By Lemma 40, and the fact that π is open, $\pi(K)$ is a neighborhood of 1 in G/H. Let a one-parameter subgroup Y in G/H be given. We can assume that $Y(t) \epsilon \pi(K) \subset U_1$ for $|t| \leq 1$. There exists $x \epsilon K$ with $\pi(x) = Y(1)$, and $X \epsilon K_1$ with $X(1) = x$. We claim that X maps on Y. If we write Z for the image of X, we have that $Y(1) = Z(1)$ and that $Y(t)$ and $Z(t)$ lie in U_1 for $|t| \leq 1$. Uniqueness of squaring in U_1 proves that $Y = Z$.

10. Approximation by NSS groups.

The proof of the main theorems of this section will need the Peter-Weyl theorem. We state it in the following way: if G is compact, and $x \neq 1$ in G, there exists a continuous representation (i. e., a homomorphism into n by n non-singular complex matrices) which does not send x to the identity matrix. Now the group of n by n non-singular complex matrices is NSS (see §1 for a sketchy proof). This enables us to recast Peter-Weyl into the form that we shall use.

Lemma 41. Any neighborhood of 1 in a compact group G contains a closed normal subgroup N such that G/N is NSS.

Proof. Let V be the given neighborhood of 1; take V open. Write $\{S_\alpha\}$ for the set of kernels of continuous matrix representations of G. The Peter-Weyl theorem says that $\bigcap S_\alpha = 1$. If V' is the (closed) complement of V, $\bigcap (S_\alpha \cap V')$ is void. By compactness, a suitable finite intersection

$$S_1 \cap S_2 \cap \ldots \cap S_n \cap V'$$

is void. Take $N = S_1 \cap S_2 \cap \ldots \cap S_n$. Then N is closed, normal, $N \subset V$, and G/N admits a faithful continuous matrix representation, to wit the direct sum of the representations with kernels $S_1, \ldots, S_n$. Thus G/N is NSS.

The machinery of §5 is now going to be invoked for the last time. For convenience of reference we make a definition.

Definition. Let G be a metrizable locally compact group. A Yamabe system (Q_i, U) in G consists of a compact symmetric neighborhood U of 1, and a sequence $\{Q_i\}$ of symmetric sets with the following three properties:

(a) $1 \in Q_i$ for all i,

(b) The subgroup generated by Q_i is not contained in U,

(c) $\bigcup Q_i$ (the set-theoretic union) has compact closure.

Two differences should be noted, as compared with the setup used for NSS groups: (1) the neighborhood U has no special properties; and moreover we shall change it at our convenience; (2) the Q_i's are not constructed from U.

The reader should now remind himself of the objects constructed in §5 from a Yamabe system: $n_i, \Delta_i, \emptyset_i$ (relative to a fixed choice of θ), and ψ_i.

Definition. $S(Q_i, U)$ is the set of all x such that there exists a sequence $y_i \in G$ with $y_i \to x$, $\Delta_i(y_i) \to 0$.

Tiny changes in the proof of Lemma 13 show that $S(Q_i, U)$ is a subgroup contained in U.

Definition. A Yamabe system (Q_i, U) is proper if $S(Q_i, U)$ is contained in the interior of U.

In Lemma 43 we shall see that we can pass from a given Yamabe system (Q_i, U) to a proper one by suitably shrinking U. However, in order to look after another property still to be discussed, we need the freedom to do this in a certain way. We prove Lemma 42 as a prelude.

Lemma 42. Let U be a compact symmetric neighborhood of 1 in a metrizable locally compact group. There exists a sequence $\{W_i\}$ of open symmetric neighborhoods of 1 with the following property: given $N \subset V \subset U$ with N compact symmetric, $1 \in N$, and V symmetric open, there is a W_j satisfying

$$N \subset W_j \subset \overline{W}_j \subset V .$$

(The bar denotes closure.)

Proof. U is compact metric, and hence has a countable dense subset $u_1, u_2, \cdots$. We take for the W_i's all finite unions of all open balls obtained by using all rational radii and centers the u's, each union made symmetric by intersecting with its inverse (the W_i's can be numbered off in any fashion). To see that our choice of $\{W_i\}$ fulfils the requirement, let N and V be given as in the hypothesis. Fix $x \in N$ for the moment. We can find one of the u's, say u_k, and a positive rational number r, such that the following two things hold: $\rho(u_k, x) < r$ (where ρ denotes the metric), and the open ball with center u_k and radius r has its closure in V. Now vary x, and cover N with a finite number of these open balls. Make this union symmetric by intersecting it with its inverse. The result is a W_j of the required kind.

Lemma 43. Let (Q_i, U) be a Yamabe system. Then for a suitable W_j of Lemma 42, $(Q_i, \overline{W}_j)$ is a proper Yamabe system.

Proof. Let us simply write S for $S(Q_i, U)$. Let U_o be the intersection of $\overline{S}$ and the interior of U. Then U_o is an open neighborhood of 1 in the compact group $\overline{S}$. By Lemma 41, U_o contains a closed normal subgroup N of $\overline{S}$ such that $\overline{S}/N$ is NSS. Lift back to $\overline{S}$ a symmetric open neighborhood of 1 in $\overline{S}/N$ that contains no subgroup $\neq 1$; the result is a symmetric open set V_1 in $\overline{S}$ such that $V_1 \supset N$ and all subgroups contained in V_1 are contained in N. We can moreover arrange $V_1 \subset U_o$ by intersecting with U_o. There exists an open symmetric neighborhood V of 1 in G with $V \cap \overline{S} = V_1$; we can moreover arrange $V \subset U$ by intersecting with the interior of U.

Now we use Lemma 42 to pick W_j with $N \subset W_j \subset \overline{W}_j \subset V$. We claim that the Yamabe system $(Q_i, \overline{W}_j)$ is proper. Let n_i' and Δ_i' be the objects formed for $(Q_i, \overline{W}_j)$. Then clearly $n_i' \leq n_i$, whence $\Delta_i' \geq \Delta_i$. It follows that $S' = S(Q_i, \overline{W}_j)$ is contained in S. Also $S' \subset \overline{W}_j \subset V$, so that $S' \subset \overline{S} \cap V = V_1$. Since every subgroup contained in V_1 is contained in N, we deduce that $S' \subset N$. Now $N \subset W_j$, and W_j is an open subset of $\overline{W}_j$. It follows that S' is contained in the interior of $\overline{W}_j$, as required.

We introduce an elaborate definition, designed to meet our needs in proving Theorem 17.

Definition. The Yamabe system (Q_i, U) is <u>tight</u> if the following three conditions hold:

(1) $\emptyset_i$ converges,

(2) There exists elements $c_i \in Q_i^{n_i}$, $c_i \notin U$, such that c_i converges,

(3) $n_i(a_i \psi_i - \psi_i)$ converges, where $a_i \in Q_i$ is one of the n_i factors of c_i, chosen so as to maximize $\|a_i \psi_i - \psi_i\|$ among these factors.

Lemma 44. If (Q_i, U) is a Yamabe system, it is possible to pick a subsequence of the Q_i's so as to get a tight Yamabe system.

Proof. We observe that $Q_i^{n_i} = Q_i Q_i^{n_i - 1}$ where $Q_i^{n_i - 1} \subset U$, and we recall that, in a Yamabe system, $\bigcup Q_i$ has compact closure. So, for every i, $Q_i^{n_i}$ is contained in the compact set

$$\overline{(\bigcup Q_i)} U \ .$$

Thus after any sequence $c_i \in Q_i^{n_i}$ has been selected, it is possible to pass to a convergent subsequence of $\{c_i\}$. Reference to Lemmas 7 and 10 then shows how to complete the job: three times in succession we pass to a subsequence.

Lemma 45.   Let (Q_i, U) be a Yamabe system. By taking a subsequence of the Q's and suitably shrinking U we can obtain a Yamabe system which is both proper and tight.

Proof.   The proof uses Cantor's diagonal procedure. With the sets $W_1, W_2, \ldots$ selected as in Lemma 42, we pass to the Yamabe system $(Q_i, \overline{W}_1)$. By Lemma 44, we can achieve tightness by passing to a subsequence of $\{Q_i\}$. Using a double index notation, we write $Q_{11}, Q_{12}, \ldots$ for the sequence in question. Thus we have that the Yamabe system

$$(Q_{11}, Q_{12}, \ldots, Q_{1n}, \ldots; \overline{W}_1)$$

is tight. We drop to $\overline{W}_2$, and pass to a subsequence again. Then the Yamabe system

$$(Q_{21}, Q_{22}, \ldots, Q_{2n}, \ldots; \overline{W}_2)$$

is tight. We continue in this way, and consider the diagonal. For the Yamabe system (Q_{ii}, U) it is the case by Lemma 43 that some $(Q_{ii}, \overline{W}_j)$ is proper. Evidently $(Q_{ii}, \overline{W}_j)$ is both proper and tight.

We are ready for the first main theorem of this section.

Theorem 17.   Let U be a compact symmetric neighborhood of 1 in a metrizable locally compact group G. Then there exists a neighborhood V of 1 such that the subgroup generated by all subgroups in V lies in U.

Proof.   Assume the contrary. Pick a fundamental system of neighborhoods of 1:

$$V_1 \supset V_2 \supset \ldots \supset V_n \supset \ldots$$

We can assume $\overline{V}_1$ compact. Define Q_i to be the set-theoretic union of all subgroups in V_i. Then (Q_i, U) is a Yamabe system (observe that our denial of the conclusion of the theorem assures us that, for each each i, the subgroup generated by Q_i is not contained in U). By Lemma 45, by passing to a subsequence and shrinking U, we can assume that (Q_i, U) is both proper and tight (change notation for both Q_i and U). We are ready for the machinery of §5. We have $\phi_i \to \phi$, so that $\psi_i \to \psi$ (Lemma 11). Write $\gamma_i = n_i(a_i\psi_i - \psi_i)$, and say $\gamma_i \to \tau$. We have

$$
(30) \qquad a_i^{n_i}\psi_i - \psi_i = \frac{1}{n_i}\left[(\gamma_i - \tau) + a_i(\gamma_i - \tau) + \ldots + a_i^{n_i-1}(\gamma_i - \tau)\right]
$$
$$
+ \frac{1}{n_i}\left[(\tau - \tau) + (a_i\tau - \tau) + \ldots + (a_i^{n_i-1}\tau - \tau)\right] + \tau.
$$

Let $i \to \infty$ in (30). Observe that all powers of a_i lie in V_i. This tells us that the left side of (30) approaches 0, while the right side approaches τ. Hence $\tau = 0$. That is,

$$
\| n_i(a_i\psi_i - \psi_i) \| \to 0.
$$

Since

$$
\| c_i\psi_i - \psi_i \| \leq n_i \| a_i\psi_i - \psi_i \|
$$

we deduce that $c_i\psi_i - \psi_i \to 0$, whence $c\psi = \psi$, where $c = \lim c_i$. Therefore $\psi(c) = \psi(1)$. This means that when the Schwartz inequality is invoked for

$$
\psi(c) = \int \phi(cy)\phi(y)\,dy
$$

we get equality, which is possible only if $\phi(cy)$ is a scalar multiple of $\phi(y)$, $\phi(cy) = \lambda\phi(y)$. Putting $y = 1$, we get $\phi(c) = \lambda\phi(1) = \lambda$, so that $\lambda \leq 1$. Putting $y = c^{-1}$ we get $1 = \lambda\phi(c^{-1})$, so that $\lambda \geq 1$. Hence $\lambda = 1$. We have proved that $\phi(c) = 1$. Hence (Lemma 12), $c \in S$, where S is again the subgroup $S(Q_i, U)$ attached to the Yamabe system

(Q_i, U). But S is contained in the interior of U, while $c = \lim c_i$ with $c_i \notin U$. This contradiction completes the proof of Theorem 17.

Theorem 18. Let G be a connected locally compact group, U a neighborhood of 1 in G. Then there exists a closed normal subgroup $N \subset U$ such that G/N is NSS.

Remark. Gluskov [9] has improved this theorem by assuming only that G is compact modulo its component of the identity. This is a pleasant hypothesis, since any locally compact group contains an open subgroup of this sort. Gluskov's work also enables one to delete metrizability in Theorem 17. On the other hand, some hypothesis is needed in Theorem 18, as is shown by the existence of simple non-discrete totally disconnected locally compact groups.

Proof. Theorem 1 reduces the problem to the metrizable case.

We may assume U is compact. We apply Theorem 17 to get an open neighborhood V of 1, $V \subset U$, such that the closure T of the subgroup generated by all subgroups in V lies in U. Thus T is compact. Put $W = V \cap T$, so that W is an open neighborhood of 1 in T. By Lemma 41, W contains a closed normal subgroup N of T such that T/N is NSS. Pick an open neighborhood of 1 in T/N containing no subgroup $\neq 1$, and lift it back to an open neighborhood W_1 of 1 in T; by intersecting with W we can arrange that $W_1 \subset W$. We have the property that all subgroups of T contained in W_1 lie in N. Take an open neighborhood X of 1 in G, $X \subset V$, $X \cap T = W_1$. Every subgroup contained in X lies in V, hence in T, hence in $X \cap T = W_1$, and hence in N. By continuity and a finite covering argument, there exists a neighborhood Y of 1 such that $aNa^{-1} \subset X$ for $a \in Y$. Now aNa^{-1} is a subgroup lying in X; therefore $aNa^{-1} \subset N$. The set H of all a with $aNa^{-1} = N$ is a subgroup of G which is open since it contains $Y \cap Y^{-1}$. By connectedness, H = G and N is normal in G. Theorem 18 is proved.

11. Further developments.

From the point which the theory has now reached, the procedure is quite standard for continuing until the subject merges with classical Lie group theory. This was well understood before the big breakthrough of 1952. In this final section we shall therefore confine ourselves to a sketch, in a series of numbered comments, of the way the course continued in the Winter of 1958.

1. Let G be a connected locally compact group. By Lemma 41 there exists a family $\{N_i\}$ of closed normal subgroups in G such that $\bigcap N_i = 1$ and each G/N_i is NSS. Thus we have an isomorphism of G into the Cartesian product $\prod G/N_i$. It is a homeomorphism, and the image is closed. In this way we get a description of G as a closed subgroup of a product of NSS locally compact groups.

2. From this information we deduce the existence of a vector space structure on the one-parameter subgroups of G. In all essential respects this vector space behaves as it does in the NSS case. We again use the notation $L(G)$.

3. (a) Let H be a closed normal subgroup of G such that G/H is NSS. Then one-parameter subgroups of G/H can be lifted to G. To see this, we take a set $\{N_i\}$ of closed normal subgroups as in paragraph 1. Each $G/(H \cap N_i)$ is NSS. We lift the appropriate one-parameter subgroup from $G/(H \cap N_i)$ to G/N_i by Theorem 16. An inverse limit argument finishes the job.

(b) Thus the natural map from $L(G)$ to $L(G/H)$ is onto. We can map $L(G/H)$ linearly back into $L(G)$, in such a way that the composite is the identity. If n is the dimension of $L(G/H)$, this yields a local lifting from G/H to G of an n-cell.

4. To continue the discussion with the use of a minimal amount of topology, we make an ad hoc definition: a topological space is feebly finite-dimensional if, for some n, it does not contain a homeomorphic copy of an n-cell. If our group G is feebly finite-dimensional, then there is a fixed upper bound to the dimensions of the NSS homomorphic images G/H.

5. Theorem: if G is a connected locally compact group, and G/H has maximal dimension among all NSS homomorphic images of G, then H is totally disconnected. Sketch of the proof: pick $\{N_i\}$ as above. The map

$$L(G/(H \cap N_i)) \to L(G/H)$$

is onto. Maximality of the dimension of $L(G/H)$ forces it to be one-to-one. Therefore the kernel is an NSS group containing no one-parameter subgroups, and is discrete (Theorem 6). The component of the identity H_o of H maps into 1 in each $H/(N_i \cap H)$. Hence $H_o \subseteq \bigcap N_i = 1$.

6. Suppose in addition that G is locally connected. Then H is discrete and G is NSS. This follows quickly from examining the image in G/H of a connected neighborhood of 1 in G, selected small enough so that its image falls into the neighborhood of 1 in G/H ruled by one-parameter subgroups.

7. We summarize, incorporating the information that if G is locally connected we need not assume G to be connected: if G is locally compact, locally connected, and feebly finite-dimensional, then G is NSS. Corollary: G is locally Euclidean if and only if G is NSS.

8. At last we can show that the NSS condition is preserved under homomorphic image. One does this by noting that G/H inherits from G its local connectedness and the bound on the dimension of its NSS images.

At this point we might be bold enough to drop "NSS locally compact" in favor of "Lie".

9. Let G be a Lie group, L its vector space of one-parameter subgroups. We put an algebra structure on L. For $a \in G$, write S_a for the linear transformation on L induced by $x \to axa^{-1}$. If $F(L)$ is the full linear group on L, we have a continuous homomorphism: $G \to F(L)$. It sends $X \in L$ into a one-parameter subgroup in $F(L)$. Now elementary arguments determine the structure of the one-parameter subgroups in $F(L)$: each has the form $t \to e^{At}$ for a suitable linear transformation A on L. For the A that comes from X we write $A = \Lambda(X)$. Thus

$$S_{X(t)} = e^{\Lambda(X)t} \quad .$$

We define $[XY] = \Lambda(X)Y$. It is routine to see that we have put on L the structure of an anti-commutative algebra. We write $\mathcal{G}$ for this algebra, and use an analogous notation for any Lie group.

10. Next comes the proof that

$$\lim_{i \to \infty} \lim_{j \to \infty} \{X(\tfrac{1}{i})Y(\tfrac{1}{j})X(-\tfrac{1}{i})Y(-\tfrac{1}{j})\}^{[rij]}$$

exists and equals $[XY](r)$ for any $r \geq 0$.

11. A corollary of this formula is that a continuous homomorphism from a Lie group G into a Lie group H induces an algebra homomorphism from $\mathcal{G}$ into $\mathcal{H}$.

12. In particular $X \to \Lambda(X)$ preserves the bracket operation. This is the Jacobi identity. So $\mathcal{G}$ is a Lie algebra.

13. Theorem: let H be a closed normal subgroup in a Lie group G. Then $\mathcal{H}$ is an ideal in $\mathcal{G}$, and $\mathcal{G}/\mathcal{H}$ is in a natural way the Lie algebra of G/H.

14. Call a subgroup S of a Lie group G <u>analytic</u> if it is generated by the one-parameter subgroups it contains and they form a Lie subalgebra $\mathcal{P}$ of $\mathcal{G}$. There is a one-to-one correspondence between analytic subgroups of G and Lie subalgebras of $\mathcal{G}$. An analytic subgroup S need not be closed. However, it can be retopologized to be a Lie group. More exactly, there exists a connected Lie group S_o with Lie algebra $\mathcal{P}$ and a continuous isomorphism of S_o onto S.

15. In a suitable neighborhood of 0 in $\mathcal{OJ}$ we can define a multiplication by transferring the group operation. We write $\times$ for this product, so that $(X \times Y)(1) = X(1)Y(1)$.

Define $\Gamma(X) = \int_0^1 S_{X(t)}\, dt$. Γ is continuous and $\Gamma(0) = I$. Hence Γ is non-singular in a neighborhood of 0, and we may define $\Delta = \Gamma^{-1}$ near there. Theorem:

$$(31) \qquad \frac{d}{dt}(X \times tZ) = (\Delta(X \times tZ))Z \ .$$

The proof is lengthy, involving the use of the integral formula proved in Lemma 37, standard sequences, and differentiable functions in the style of §9. In this way it is first verified that $(-X) \times (X + tY)$ is differentiable at $t = 0$ with derivative $\Gamma(-X)Y$. Brouwer's theorem on invariance of domain is invoked to invert this result and prove (31).

Equation (31) is a differential equation showing that the group operation is determined locally by the Lie algebra. By standard theorems on differential equations, (31) can be used to prove that $X \times Y$ is analytic, and that a Lie algebra homomorphism from $\mathcal{OJ}$ to $\mathcal{L}$ preserves $\times$ in a neighborhood of 0.

BIBLIOGRAPHY

1. R. Block, On Lie algebras of classical type, Proc. Amer. Math. Soc. 11 (1960), 377-9.

2. _____, Trace forms on Lie algebras, Can. J. of Math. 14(1962), 553-554.

3. _____, The Lie algebras with a quotient trace form, Ill. J. of Math. 9(1965), 277-285.

4. R. Block and H. Zassenhaus, The Lie algebras with a non-degenerate trace form, Ill. J. of Math. 8(1964), 543-9.

5. N. Bourbaki, Groupes et Algèbres de Lie, Hermann, Paris. Ch. I, 1960, Chs. II, III, 1972, Chs. IV, V, VI, 1969.

6. C. Chevalley, Théorie des groupes de Lie, Tome II: Groupes Algébriques, Hermann, Paris, 1951.

7. I. M. Gelfand, Zur Theorie der Charaktere der Abelschen topologischen Gruppen, Mat. Sbornik 9(1941), 49-50.

8. A. Gleason, Groups without small subgroups, Ann. of Math. 56(1952), 193-212.

9. V. M. Gluskov, The structure of locally compact groups and Hilbert's fifth problem, Uspehi Mat. Nauk 12(1957), no. 2, 3-41. AMS Translation 15(1960), 55-93.

10. R. Jacoby, Some theorems on the structure of locally compact local groups, Ann. of Math. 66(1957), 36-69.

11. N. Jacobson, Lie Algebras, Interscience, 1962.

12. I. Kaplansky, Lie algebras of characteristic p, Trans. Amer. Math. Society 89(1958), 149-183.

13. V.A. Kreknin, Solvability of Lie algebras with a regular automorphism of finite period, Dokl. Akad. Nauk SSR 150(1963), 467-9.

14. L. Loomis, An Introduction to Abstract Harmonic Analysis, Van Nostrand, 1953.

15. W. Mills, Classical type Lie algebras of characteristic 5 and 7, J. Math. Mech. 6(1957), 559-566.

16. D. Montgomery and L. Zippin, Small subgroups of finite-dimensional groups, Ann. of Math. 56(1952), 213-241.

17. _____, Topological Transformation Groups, Interscience, New York, 1955 (1965 reprinting has additional bibliography).

18. J. R. Schue, Hilbert space methods in the theory of Lie algebras, Trans. Amer. Math. Soc. 95(1960), 69-80.

19. G. Seligman, On Lie algebras of prime characteristic, Memoirs Amer. Math. Soc., no.19,1956.

20. _____, Some remarks on classical Lie algebras, J. Math. Mech. 6(1957), 549-558.

21. _____, Modular Lie algebras, Ergebnisse der Math. vol.40, Springer, 1967.

22. G. Seligman and W. Mills, Lie algebras of classical type, J. Math. Mech. 6(1957), 519-548.

23. Seminaire Sophus Lie, Ann. Ecole Norm. Sup. 1954-5.

24. J.-P. Serre, Algèbres de Lie Semi-simples et Complexes, Benjamin, 1967.

25. J. R. Shoenfield, The structure of locally compact groups, Duke University, 1956 (mimeographed). Reviewed in Mathematical Reviews 18(1957), 317.

26. D. J. Winter, On groups of automorphisms of Lie algebras, J. of Alg. 8(1968), 131-142.

27. H. Yamabe, On the conjecture of Iwasawa and Gleason, Ann. of Math. 58(1953), 48-54.

28. _____, A generalization of a theorem of Gleason, Ann. of Math. 58(1953), 351-365.

29. H. Zassenhaus, On trace bilinear forms on Lie algebras, Proc. Glasgow Math. Assoc. 4(1959), 62-72.

ADDITIONAL BIBLIOGRAPHY (ADDED 1974)

On Lie algebras

James E. Humphreys, Introduction to Lie Algebras and Representation Theory, Springer, 1972.

David J. Winter, Abstract Lie Algebras, MIT Press, 1972.

On the fifth problem

J. Dieudonné, Eléments d'Analyse, vol. 4, Gauthier-Villars, 1971. (On pages 162-166 the proof is outlined in five brisk exercises.)

P. Enflo, Topological groups in which multiplication on one side is differentiable or linear, Math. Scand. 24(1969), 195-207.

P. Enflo, Uniform structures and square roots in topological groups, Israel J. Math. 8(1970), 230-252 and 253-272. (Enflo's papers invade the infinite-dimensional case.)

S. N. Hudson, Lie loops with invariant uniformities, Trans. Amer. Math. Soc. 115(1965), 417-432 and 118(1965), 526-533. (Hudson is non-associative.)

B. A. Pasynkov, On topological groups, Dokl. Akad. Nauk SSSR 118(1969), 286-289. AMS translation 10(1969), 1115-1118. (The locally pseudo-compact case. "Pseudo-compact" means that every real continuous function is bounded.)

E. G. Sklyaryenko did the writup on the fifth problem in the 1969 Russian book on Hilbert's problems (German translation, Akad. Verlag., Leipzig, 1971).

INDEX

Abelian, 5

Algebra, 1

Canonical neighborhood, 93

Cartan subalgebra, 18, 43

Center, 5

Centroid, 30

Commutation, 2

C-system, 54

Derivation, 4

    inner, 4

Differentiable, 98

D-system, 57

Engel's theorem, 14

Equicontinuous, 99

Form, 27

    invariant, 27

    Killing, 35

Haar measure, 103

Ideal, 9

Jacobi identity, 1

Length, 10, 13

Level, 60

Lie

    algebra, 1

    group 4, 87, 140

    ring, 1

    set, 15

Lie's theorem, 20, 24

Locally Euclidean, 87

Modulus, 96

Nicely embedded, 57

Nil, 12

Nilpotent, 12

Normalizer, 42, 47

NSS (no small subgroups), 87

One-parameter subgroup, 92

Peter-Weyl theorem, 131

Proper

    function, 102

    Yamabe system, 132

Radical, 11

Rank 43, 75

Regular, 43

Representation, 17

    adjoint, 18

    completely reducible, 18

    faithful, 18

    irreducible, 18

    kernel of, 18

    projective, 67

    regular, 18

Restricted, 73

Ring, 1, 9

26

Root, 41, 53, 75

    isotropic, 80

    space, 41

Semi-simple, 11

Series

    derived, 10

    descending central, 12

Simple

    algebra, 9

    root, 58

Solvable, 10

Standard, 96

Symmetric, 89

Tight, 134

Translation, 98

V-algebra, 75

Witt algebra, 4

Yamabe system, 131

    proper, 132